THE WORLDS OF ANTS, BEES AND WASPS

Also by Brian Vesey-FitzGerald

THE WORLD OF FISHES

THE WORLD OF REPTILES

ENQUIRE WITHIN ABOUT ANIMALS

THE CAT OWNER'S ENCYCLOPAEDIA

THE DOG OWNER'S ENCYCLOPAEDIA

BRITISH GAME

PORTRAIT OF THE NEW FOREST

THE VANISHING WILD LIFE OF BRITAIN

etc.

THE WORLDS OF ANTS, BEES AND WASPS

Brian Vesey-FitzGerald

PELHAM BOOKS

First published in Great Britain by
PELHAM BOOKS LTD
26 Bloomsbury Street
London, W.C.1
1969

© 1969 by Brian Vesey-FitzGerald

All Rights Reserved. No part of this publication may be reproduced, stored in a retrieval system, or transmitted, in any form or by any means, electronic, mechanical, photocopying, recording or otherwise, without the prior permission of the Copyright owner

7207 0252 6

Set and printed in Great Britain by Tonbridge Printers Ltd, Peach Hall Works, Tonbridge, Kent, in Bembo twelve on fourteen point, on paper supplied by P. F. Bingham Ltd, Croydon, and bound by James Burn at Esher, Surrey

CONTENTS

1	The Wonderful Insect	*page* 11
2	Woodborers and Gall Causers	25
3	The Parasites	40
4	The Solitaries	54
5	Social Bees and Wasps	68
6	The Top of the Class	82
7	Their World and Ours	97
	Suggested further Reading	113
	Index	115

ILLUSTRATIONS

Between pages 24 & 25

Larvae of the sawfly *Croesus septenrionalis* on birch

Bedeguar Galls, commonly known as Robin's Pincushions, are caused by the gall-wasp *Rhoditus* (syn. *Diplolepis*) *rosae*.

Oak Apples are caused by the gall-wasp *Biorrhiza pallida*, the unisexual generation of which causes gall-formation on the roots of oaks

Marble Galls are caused by the gall-wasp *Cynips kollari*, the life-history of which is still imperfectly known

Between pages 40 & 41

Larvae of the Pine Sawfly (*Lophyrus pini*)

All too familiar to the gardener: larvae of the Gooseberry Sawfly

Female ichneumon *Rhyssa persuasoria*

Dead caterpillar of the Scarce Vapourer Moth surrounded by the fully fed larvae of the Apanteles that killed it

Between pages 72 & 73

Female Sand Wasp dragging a paralysed sawfly larva to her nest

Female Leaf-cutter Bee on rose leaf

Female Mining Bee
Female Bumblebee (*Bombus agrorum*) on sallow catkins

Between pages 88 & 89
Worker Honeybee on daisy
Queen Wasp
Head of Queen Hornet . . . greatly enlarged
Garden Black Ants (*Lasius niger*) with cocoons

Photographs by George E. Hyde F.R.E.S.

ACKNOWLEDGMENTS

My grateful thanks are due to Oxford University Press for permission to quote a passage from *An Insect Book for the Pocket* by Edmund Sandars and to Methuen & Co. Ltd for a similar courtesy with regard to a passage from *The Senses of Insects* by H. Eltringham, F.R.S.

CHAPTER ONE

THE WONDERFUL INSECT

The ants, the bees and the wasps are, as everyone knows, insects. Not all laymen, however, fully appreciate what extraordinary creatures insects, all insects, are. Compared with vertebrates – fantastic though this may seem – insects are inside-out and upside-down animals.

In a vertebrate animal the skeleton is an internal framework around which the organs and tissues are arranged. It is exactly the other way round with the insect. In an insect the skeleton is the external shell to the inner surface of which the organs and tissues are attached. Insects are, in very truth, inside-out animals.

In a vertebrate animal the heart lies below or in front of the alimentary canal. In an insect the heart lies above or behind the alimentary canal. In a vertebrate animal the central nervous system runs along the back: in an insect it runs along the underside of the body. Insects are, in very truth, upside-down animals.

And the insects are fantastic in innumerable other ways. Organs often seem to be located with a total disregard for reason. For example, grasshoppers and certain moths have their 'ears' on the base of the abdomen just behind the wings whereas katydids and crickets, though closely related to the grasshoppers, have their 'ears' on the front legs just below the knees. Again, the same organ may perform markedly different functions in different species. Yet again, in order to perform some particular function an organ will sometimes undergo a radical change and then, the function which demanded the change having been performed, revert to its original purpose. And when it comes to reproducing their like, the insects are away out on their own. They have thought of everything, absolutely everything: most of it inconceivable. But then it has been said: 'Name the inconceivable and we will find it in the life of some insect.'

Now, there are enormous numbers of insects: not only enormous numbers of individuals, but enormous numbers of species. About 850,000 different species have already been identified and named, and new species are being discovered and named at the rate of several hundred a year. As insects are present everywhere throughout the world wherever plants grow and as there are not all that many entomologists and collectors, it may well be that man will never know exactly how many different species there are. Conservative entomologists are of the opinion that there are probably well over a million. One American Professor of Entomology has suggested that there may be as many as ten million! Obviously, when you are talking in terms of a million or more *different* species, it is impossible to speak of a *typical* insect. Nevertheless, there are a number of common factors which condition the physiological make-up of the insects, which serve to link them together in one class and to distinguish them from all the members of all the other classes which form the phylum *Arthropoda;* the phylum of animals with segmented bodies, jointed legs, and an outer skeleton. All insects have three body regions, all have three pairs of walking legs when adult and many then have wings, all have one pair of true antennae, and all breathe by means of tracheae. Moreover, insects are essentially terrestrial animals.

It is this which determines the character of the cuticular skeleton, which in turn conditions the respiratory mechanism and the physiology of growth, restricting size. All insects are small. Few of the largest are as big as the smallest bird: many are so small as to be barely visible to the naked eye.

The outer skeleton is composed of three layers of tough material forming a coating which not only covers the whole of the body and its appendages, such as the legs, but also turns inward at the mouth and the anus to line the front and rear portions of the alimentary canal, and at the spiracles to line the breathing tubes (the *tracheae*). This outer skeleton (the exoskeleton) has two principal functions: to protect and support the soft parts of the body, and to provide an adequate foundation for the muscles. For the accomplishment of these two functions it must be more or less thick and hard: at least in part. (If it were hard throughout, it would be more of a handicap than an advantage to its owner, for it would make movement extremely difficult.) The best known constituent of the exoskeleton is a substance called *chitin*, which is present only in the inner two layers. Chitin is a tough, fibrous, flexible material which plays the same part in the insects as does the layer of slime or mucus in the slugs and the snails. Chitin may, indeed, properly be thought of as a kind of tough dry mucus. But chitin, though tough, is not hard. The thickened hardened parts of the exoskeleton are made up of a horny material called *sclerotin*, which, basically, is not unlike *keratin*, the horny substance of which the scales of reptiles, the feathers of birds, and the hoofs and horns of mammals are composed. Sclerotin is light, but extremely strong: chitin is tough and flexible. Every advantage is taken of the contrast between the two, of the combination of the two. Chitin provides the hinges and joints: sclerotin the armour plating. The exoskeleton, therefore, consists of a number of rigid hard plates, known as *sclerites*, which are separated from each other (and joined to each other) by bands of tough flexible material. All insects have two such tough flexible bands completely encircling them in such a way as to divide their bodies into three regions: the head, the thorax, and the abdomen. The thorax and the abdomen are subdivided into a number of rings or segments, which are themselves divided into sclerites by criss-cross lines of flexible material. And these bands of flexible material occur wherever there is an appendage and at each of its joints.

The head, except for its appendages, is not clearly divided into segments. The appendages are the antennae and the mouth-parts. The mouth-parts are subject to great variation in form and function, according to the nature of the food, but fall into two main groups: chewing or biting mouths and sucking mouths. A knowledge of the general type of mouth-parts found in different groups of insects is of great help in understanding their habits and, for the expert, also in classification. In the typical chewing or biting mouth the *labrum* or upper lip projects from the front of the head as a kind of flap. Immediately behind the labrum are the *mandibles* or principal jaws. These are large, hard, strong, toothed jaws which work from side to side instead of up and down as do human jaws. Behind the mandibles is a second pair of jaws, the *maxillae*. Each maxilla bears an outer and an inner lobe, termed the *galea* and the *lacinia* respectively. At the base of the galea there is a sensory appendage or maxillary palp. Finally, behind the maxillae, is the *labium* or lower lip. The labium is, in fact, a second pair of maxillae which have become fused together on their inner margins. The free margin may bear as many as four lobes, together with a pair of labial palpi which have a sensory function.

The sucking mouth is much more complex and its mechanism much more difficult to understand. Many forms of insect life subsist entirely on a fluid diet and their mouth parts are modified accordingly. Everything depends upon the type of fluid: the mouth parts of butterflies and moths are quite different from those of mosquitoes and fleas, which are blood-suckers, and these, in turn, from those of greenfly, which are sap-suckers. In some insects, by the way, the mouth is altogether useless.

The antennae also vary enormously in different species, but mostly they resemble jointed threads or strings of beads. The joints or beads are the segments of the antennae. The antennae of nearly all, probably all, insects are sensitive to touch and so they may accurately be called 'feelers'. But in many insects the antennae, while undoubtedly serving as feelers, perform other functions as well. In male mosquitoes they seem also to function as hearing organs. In the males of many moths, and in various ants, bees, wasps, flies and beetles they seem also to serve as organs of smell.

The head also includes the eyes. If that seems to be a statement of the obvious and therefore unnecessary, remember the strange places where the 'ears' of some insects are situated. Insects have two kinds of eyes: compound and simple. The compound eyes are two in number, are large and usually round or oval, and are situated on the sides of the head. When examined under a microscope the surface of each compound eye appears to be divided into a number of hexagonal areas, which are known as *facets*. These facets are the outer ends of the separate lenses (the *ommatidia*) of which the compound eyes are composed. Each lens is connected with its own independent nerve and carries to the brain an image of what is directly before it. This means that the brain of an insect receives a number of different pictures at the same time – the picture recorded by each individual facet – and pieces them together to make one whole. This is the principle of the mosaic created by the human artist: a great many small pieces of material which individually are meaningless, but which when pieced together make a brilliant and meaningful picture. The compound eyes of most ground-dwelling insects are composed of only a few, comparatively large, facets. The compound eyes of dragonflies and certain wasps, on the other hand, are made up of many thousands of facets. Clarity of vision in insects is said to be roughly proportionate to the number of facets in the compound eye. At least there can be no doubt whatever that the dragonflies, which hunt their prey and seek their mates by sight, have exceptionally keen eyesight.

The simple eyes (*ocelli*) are usually three in number and are small bead-like organs which are set on the top or on the front of the head between the compound eyes. Remarkably little, indeed virtually nothing, is known about the function of the simple eyes. And the fact that while they are found in many species they are absent in numerous others does not make understanding of their function any the easier. They are, for example, hardly ever present in the beetles whereas they are present in all dragonflies. They are present in many species of flies and absent in others. And in some insects – moths and bees, for example – while present they are wholly or partially concealed by scales or hairs. The simple eyes are generally dismissed as inefficient visual organs, capable of no more than distinction between light and dark. But the enormous diversity of structure in simple eyes implies a wider range of function and efficiency than that. It is difficult to believe that the very complex *ocelli* of certain dragonflies can do no more than distinguish between light and dark: a function already adequately performed by their very efficient compound eyes. It has been suggested that the more complex, more highly developed, simple eyes may enable the insect to recognise objects at very close range. It is not explained why the insect, why a dragonfly for example, should need to do so when it is gifted by the presence of two very efficient compound eyes which would already have recognised the object at a distance. Moreover, the absence of any focal mechanism in the *ocelli* suggests that recognition would be uncertain and unlikely. It may be that the *ocelli* should not be considered as eyes in the strict sense, but rather as stimulators of the compound eyes. It may well be that their function is to quicken the response of the insect to changes in the brightness of light as seen by the compound eyes. It has been shown that if the simple eyes are covered the insect's reaction to light is much less positive.

The thorax is almost always divided into three distinct segments, one behind the other. These segments are themselves divided horizontally, so that each is made up of six sclerites: one on top, one underneath, and two on each side. The thoracic segments support the legs and, if there are wings, the wings. In all insects the first segment (the *prothorax*) supports the front legs, the second segment (the *mesothorax*) the middle pair of legs, the third segment (the *metathorax*) the hind legs. That is an invariable rule: there are no exceptions. And in all insects, if there are wings, the fore pair springs from the mesothorax above the middle pair of legs and the hind pair from the metathorax above the hind legs. That, too, is an invariable rule. The prothorax never supports wings.

Typical insects breathe not by means of lungs or gills, but by means of a system of branching air tubes, the *tracheae*, which extend throughout the body and carry air to the very tissues where it is needed. The tracheal tubes open to the outside through the spiracles, which are situated on the sides of the thorax in the flexible bands separating the segments. There are two pairs of spiracles and the openings are guarded by valves.

The six legs of insects are usually all much alike, all being built on the same plan. Each leg consists of five divisions, which are named (somewhat inappropriately) after the parts of the human leg: *coxa* (hip), *trochanter* (thigh-joint), *femur* (thigh), *tibia* (shin) and *tarsus* (foot). The coxa and trochanter are short segments by which the leg is hinged to the body. The femur and tibia are long segments containing certain of the leg muscles, though the muscles which move the leg as a whole lie in the thorax. The trochanter occasionally consists of two segments – this is the case with the dragonflies and in certain wasps, for example – but normally it and all the other parts of the leg, except the tarsus, consist of but a single segment. Most commonly the tarsus consists of five segments, but this number may be, and often is, reduced by fusion or by loss to four, three, two or even one. The last of the normal five segments has two claws with either a pad or hairs between them. But claws, pads and hairs may sometimes be missing from some or all the legs. The insect foot is almost as variable as the insect mouth.

This is true also of the wings. The wings are really thin

extensions of the exoskeleton. Until the last moult they are merely sacs containing a number of tubes. After the last moult the insect forces air into these tubes, which makes them lengthen and spread out, flattening the sacs, until finally they form a fan of stiff struts, known (incorrectly) as veins, supporting a thin membrane. The number and arrangement of the wing veins are important in the classification of insects, for in them are to be found many of the evidences of relationship upon which classification is based.

The abdomen consists of a further ten to twelve segments. By comparison with the head and the thorax the abdomen bears few appendages. In fact, most abdominal segments bear no appendages at all. At the tip of the abdomen there is a group of appendages (the *genitalia*) which function in mating and egg-laying: in the male the copulatory organs and in the female a number of egg-laying tubes (generally four or six), prong-like structures which together form an *ovipositor*, which is a device for placing the eggs when they have been laid. In many of the Hymenoptera the ovipositor has been modified into a sting. In addition to the genitalia the abdomen of certain insects bears a pair of anal feelers called *cerci*. These are long and conspicuous in the Bristle-tails, but so small as to be quite inconspicuous in the grass-hoppers. Several of the abdominal segments bear spiracles, one pair to each segment, and these are usually situated on the sides. The number varies with the species, but there are never more than eight pairs.

Internally insects differ from vertebrates just as much as they do externally. It is not merely a matter of being 'upside-down' animals: everything is different. For example, the muscles instead of being tapered at the ends, as is generally the case with vertebrates, are usually strap-like and of the same width throughout. Again, instead of being bound closely to the skeleton with connective tissue, as is the case with vertebrates, the muscles extend freely across parts of the interior, somewhat on the principle of the ropes of a hammock. The tubular alimentary canal, which extends the full length of the body, is suspended within the body cavity by strands of connective tissue, by branches of the tracheae, and by muscles. The other internal organs are held in place in a similar manner. This means that around and between the various internal organs of an insect there is a great deal of space or, more correctly, there are many intercommunicating spaces.

These spaces are filled with blood. An insect's blood fills its body cavity instead of being contained in veins. The blood is pumped through each segment instead of being circulated by a local heart. And, instead of being forced through local organs in order to be aerated, it is in process of aeration all the time by means of fine tubes with thin body-walls carrying oxygen, so that wherever there is blood there is air as well. It is a very simple system and one with great advantages: the most immediate of which is that injuries which would be fatal in the vast majority of vertebrates are a matter of indifference to insects. Any serious injury to the heart or lungs of a vertebrate must result in death. But a similar injury in insects, provided only that the body fluid is not permitted to escape, will not prove fatal because the rest of the circulatory system or respiratory system will be functioning normally.

The central nervous system consists of two cords running the whole length of the underside of the body below the alimentary canal and the other organs. At intervals along their length these cords are joined and the junctions are marked by roundish swellings. These swellings are collections of nerve cells and are known as *ganglia*. There is generally one ganglion for each body segment. Tiny, thread-like nerves branch out from the ganglia and extend to the various tissues and organs, but the nerves from any one ganglion usually run only to the tissues and organs of the segment containing the ganglion. The more work a ganglion has to do the larger it is. For example, the thoracic segments are much more complex than the abdominal segments since, in addition to the usual internal tissues and organs, they have legs, and sometimes wings also, attached to them. More nerves are, therefore, needed in these thoracic segments and more cells are necessary to control their activity. The thoracic ganglia are, therefore, larger and more complex than those of the abdomen. In the head there are ganglia associated with the compound eyes (a separate ganglion for each eye), with the simple eyes, with the antennae, with the mouth-parts and with the throat. The head ganglia are usually spoken of collectively as the brain.

In mammals it is generally assumed that the convolutions of the brain are an index of mental capacity or, to use the more popular term (but one which it is extremely difficult to define), of intelligence. In general terms this does seem to be true – for example, the brains of dolphins and porpoises are very heavily convoluted and recent research has shown these animals to possess a very high level of intelligence – though it is not, of course, the whole truth. (The internal structure of the brain must be at least as important as its superficial convolutions.) On the same principle it may be presumed that it is upon the size and complexity of the brain (the head ganglia) that the instinct or intelligence of the insect depends. In general terms this is, no doubt, correct.

But it has to be remembered that the insect is a segmented creature and that each ganglion is, in fact, a sort of 'local brain' controlling the muscles of the segment in which it is situated. And there can be no doubt whatever that these local brains are capable of exercising a considerable degree of independence and even, in some cases and under certain circumstances, of taking over altogether for a period of time.

I have already pointed out that the segmented body may enable an insect to survive grievous injury. Some insects can continue to live and to function, for quite considerable periods of time, without a head. A headless moth, for example will lay her full complement of eggs. Even more surprising is the fact that a headless moth is still capable of attracting a male and that the male will mate with her without, apparently, noticing that what might be considered an essential part is missing: certainly without bothering about it. It is quite unnecessary to point out that anything comparable with these two examples cannot take place among mammals. And this sort of thing is not confined to the moths: it is common to insects generally. The severed head of a wasp continued to feed happily on marmalade; apparently unaware of the loss of its stomach. A decapitated wasp is still capable of delivering a sting. There can be no doubt that to have each segment of the body supplied with body fluid, oxygen, muscles, nerves and a 'brain' confers some very considerable advantages.

I was careful to point out at the very beginning that there is no such thing as a 'typical' insect. What I have been describing is a physical system, the segmented body, which is common to all insects. But the segmented body is a remarkably elastic type of physical structure. There is, in fact, a greater variety of form among the insects than there is in any other class of animal life. There is also, despite the characteristically small size of the class as a whole, a greater variation in size. The largest mammal in the world – and, in fact, the largest animal that has ever lived; larger by far than any of the great dinosaurs of prehistoric times – is the Blue Whale. The largest recorded specimen measured 100 ft. The smallest mammal in the world, a Pigmy Shrew, measures about two inches. The largest mammal in the world is therefore only about six hundred times as long as the smallest mammal. The largest insect in the world is a stick insect, *Pharnacia serratipes*, which measures eleven inches. The smallest insect in the world is a wasp, *Alaptus magnanimus*, which measures .21 mm or considerably less than one-hundredth of an inch. The largest insect in the world is therefore more than a thousand times as long as the smallest insect.

The small size and lightness of insects bring in their train some considerations which are not immediately apparent. For example, quite apart from powers of flight, gravity presents no terrors. If you divide the length, breadth and height of an animal by ten, you reduce its weight to one-thousandth, but its surface only to one-hundredth. As Haldane has pointed out, the resisting force to falling is therefore relatively ten times greater than the driving force. On the other hand, water presents a very real problem for the terrestrial insect. When a man steps out of his bath he carries on his body about a pound of water. A wet mouse carries its own body weight again. But a wet fly carries many times its own weight and is virtually helpless. A terrestrial insect going for a drink is running a very grave risk, for if it breaks the surface tension it is certain to drown or to become the helpless prey of some aquatic animal. This is why terrestrial insects prefer to take their liquid from wet surfaces and this is why flies are attracted to humans moist with perspiration.

But to return to variety of form. It is this ability to vary within the typical physical structure, and to do so without weakening the structure, which has enabled the insects to establish themselves firmly everywhere, even in what appear to be the most unfavourable conditions; which has enabled them to exploit successfully all this world's resources. Wherever man goes, there are the insects before him. Whatever man does, some insect has done before him: and probably done better.

You will no doubt dismiss that as an exaggeration. Well, consider man's conquest of the air. What he has accomplished is beyond measure marvellous. But his technical efficiency falls far short of that of the dragonfly, which can manipulate its four wings so that sometimes it uses the fore-pair and the hind-pair independently and sometimes interlocks them so that they act as one pair, and which can, when it so desires, reverse engines in flight and fly backwards. And nothing man has achieved in the way of aerobatics approaches the skills daily displayed by house flies just beneath the ceiling. Man explores the depths of the sea and the heights of the sky: different men in different machines. The water boatman comes from the bottom of a pond, takes off from the surface, spreads its wings and is gone: and all in the twinkling of an eye.

Larvae of the Sawfly *Croesus septenrionalis* on birch

Bedeguar Galls, commonly known as Robin's Pincushions, are caused by the gall-wasp *Rhoditus* (syn. *Diplolepis*) *rosae*

Oak Apples are caused by the gall-wasp *Biorrhiza pallida*, the unisexual generation of which causes gall-formation on the roots of oaks

Marble Galls are caused by the gall-wasp *Cynips kollari*, the life-history of which is still imperfectly known

CHAPTER TWO

WOODBORERS AND GALL CAUSERS

Of all the orders of insects, and there are some twenty-eight of them, the Hymenoptera (the order composed of the ants, bees, wasps and their relatives) is the most diverse and the most advanced. It is, however, far from being the largest in terms of species. About 275,000 species of Coleoptera (beetles) have been recognised and named, and about 200,000 species of Lepidoptera (butterflies and moths), but only about 100,000 species of Hymenoptera, which is but little more than the 85,000 known species of Diptera (the true flies). But it may be taken as certain that the present figure of 100,000 represents only a small part of the species in existence. These figures are for the whole world, of course. It is interesting to compare them with the figures for Britain. Here the order Hymenoptera heads the league table with just over 6,100 species, followed by the Diptera with just over 5,200, the Coleoptera with 3,690, and the Lepidoptera with 2,187. Here again it is certain that, even in so small an area as Britain, there are species of Hymenoptera still awaiting discovery.

The name Hymenoptera is a compound of *hymen*, a membrane, and *pteron*, a wing. The Hymenoptera have two pairs of membranous, more or less transparent wings, which are usually rather narrow and sparsely veined, the hind pair being the smaller. The wings of a side are linked together by a row of minute hooks, so that the fore-wings and the hindwings function as a single organ. In general the mouthparts are designed for biting and chewing. The mandibles are always well developed and in some species, indeed, are used for work rather than for feeding. In some species, such as the bees, an alternative method of feeding has evolved. The labium is drawn out to form a soft flexible tongue and its sides are curled downwards and inwards so as to form a tube up which nectar can be sucked.

Adult Hymenoptera are usually not very difficult to recognise because, in addition to the four clear membranous wings, a large proportion of them are extremely thin-waisted: a characteristic which will always serve to identify those members of the order (castes of ants, for example) which are wingless. Their larvae, however, are by no means so readily identified. Excepting only the larvae of saw-flies, which look very like caterpillars of butterflies or moths, and the larvae of wood-wasps or horntails, which have six stumpy 'legs' and hard heads with powerful jaws, all are legless maggots and all look alike. Often crescentic in form, all have tubular bodies of ten or eleven segments (and there is no taper as there is in the maggots of flies) and all have a well-defined head, whereas the maggots of most flies appear to be headless. Looking pretty closely, you can generally distinguish the larvae of one of the Hymenoptera from the larvae of a fly without much difficulty. But distinguishing the larvae of one species of bee from another, of one species of wasp from another, or bee larvae from wasp larvae, is a matter for the specialist: not one for such as you and I. For the amateur it is sufficient to know that the correct identification of larvae will usually be painfully impressed upon the too curious by attendant, and readily identifiable, adults!

The Hymenoptera fall naturally into two groups: the Symphyta and the Apocrita.

The Symphyta, the more primitive of the two groups, are distinguished by the fact that the thorax joins the abdomen without constriction – there is no obvious waist – and by the fact that the ovipositor has been modified to form a saw or a boring implement. The Symphyta includes the saw-flies and the wood-wasps or horntails. 'Saw-fly' is, of course, a misnomer. The true flies have only two wings – the name of their order, Diptera, means 'two wings' – but the saw-flies have the four membranous, more or less transparent, wings typical of their order. They are, in fact, primitive wasps.

In the saw-flies the ovipositor has been modified to form a two-edged saw which is used by the female to make slits or pockets, in which to place her eggs, in the leaves or stem, twigs or buds or fruits, of the plants upon which her larvae will feed. With each egg, which passes down the ovipositor between the two blades of the saw, a drop of liquid from a poison gland is extruded. This affects the tissues of the plant at that spot, preventing growth or the flow of sap, and so affording some protection to the egg. Sometimes, but not always, it causes a gall. Saw-fly eggs, after being laid, have the very unusual attribute of becoming slightly larger; presumably by absorption of liquid from the plant. Plants selected as incubators for the eggs and dining-tables for the larvae are mainly trees or shrubs, though herbaceous plants may occasionally be used.

Everyone who grows a little fruit must be familiar with the green, yellow spotted, larvae of the Gooseberry saw-fly – not only on gooseberry bushes, but also on currants, including the ornamental flowering currants of the herbaceous border – though all but a select few will believe them to be the caterpillars of some butterfly or moth. Most gardeners are not in the least interested in the identity of the insects they find upon their plants: they are concerned only with their destruction. If you should be of a more liberal and curious turn of mind, then (with the aid of a pocket lens) you will find it easy to distinguish saw-fly larvae from the caterpillars of butterflies and moths. The saw-fly larva has a large single eye set well up on each side of the head: the caterpillars of butterflies and moths have a group of six small eyes on each side of the head. Another way of distinguishing between the two is to look at the legs. In both saw-fly larva and caterpillar each of the first three segments (which correspond to the thorax of the adult insect) has a pair of jointed legs. Behind the first three segments come the abdominal segments bearing 'false legs'. Most saw-fly larvae have six or more pairs of these false legs, a pair to a segment, and there is *always* a pair on the fifth segment, which is the second abdominal segment. The caterpillars of butterflies and moths usually have five pairs of false legs and there is *never* a pair on the fifth segment. When the saw-fly larva is ready to pupate it forms a cocoon – sometimes on the food plant, but more commonly underground – and in this passes the winter. The adults are short-lived. The males die within a few hours after pairing and the females as soon as they have laid their eggs; a process which takes but a few days.

The finest members of the Symphyta group are the large and brilliantly coloured wood-wasps. The female of the giant wood-wasp, *Sirex*, is $1\frac{1}{2}$ ins long without taking into account the stout ovipositor. Many people are scared stiff of this imposing creature, being convinced beyond reassurance that it is a hornet and that the ovipositor is a particularly formidable sting. In fact, from the stinging point of view, all the wood-wasps are absolutely harmless. The stout ovipositor is used for boring holes through the bark and into the wood of conifers (broad-leaved trees are ignored): an apparently difficult operation – you try with a bradawl! – which is accomplished easily and quickly. One egg, never more, is laid in each hole and each female is believed to lay about 100 eggs. The larvae, dirty white sluggish creatures armed with very powerful jaws, spend about three years boring their way through the wood. Healthy trees, in which the sap is running freely, are very rarely attacked: but considerable damage is often caused to sickly trees and to felled timber (to which wood-wasps are particularly addicted), for the larvae may burrow deep into the heart wood. Tunnels over a foot in length have been recorded.

Do the larvae of wood-wasps eat wood? At first sight, it would seem obvious that they must. For there can, surely, be nothing else in the middle of a tree that they could eat? In fact, the larvae of wood-wasps do not eat wood. At least, they do not eat 'neat' wood. When the wood-wasps lay their eggs, they lay with each egg the conidia (the fruiting bodies) of a fungus, which forms a mould on the walls of the hole containing the egg and which subsequently spreads along the tunnels made by the burrowing larvae. Precisely what part the fungus plays in the nutrition of the larvae is not known. It is known that some woodborers are unable to digest any constituent of the wood and have to rely upon moulds for food. It may well be that the wood-wasp larva is one of these. But it may also be – and some authorities believe that it is – that the mould penetrates the surface layers of the tunnels so that they become suitable as food for the burrowing larvae. Nobody knows for certain. And nobody knows how the female wood-wasp acquires the fungus. At the base of her ovipositor there is a pair of glands which contain the conidia. But how she gets infected, so to speak, remains a mystery.

Another mystery surrounds the emergence of the adult wood-wasps from the wood that has been their home for three years. When the time comes for the larva to pupate, it forms a cocoon and lies within it with its head pointing towards the surface at its nearest point. When the adults emerge from their cocoons they have only to gnaw their way straight ahead through the wood to reach fresh air and freedom in the shortest possible time. It all sounds simple enough. But how does the larva, after three years imprisoned deep in the darkness of the timber, know in which direction the surface lies and, even more important, indeed vitally important, the shortest route to it? Most people will tell you that it knows by 'instinct'. Instinct is an umbrella word, beneath which naturalists seek shelter whenever questions become too pressing and they have not the faintest idea of the true answer. And that, in fact, is the true answer in this case: nobody knows.

All the members of the second group, the Apocrita, have a waist, often very narrow, between thorax and abdomen. In some species the waist is prolonged to form a stalk or petiole. Many members of this group – all the ants, for example, excepting only the functional males and females – have lost their wings. The Apocrita group has two sub-divisions: the Parasitica and the Aculeata.

The Parasitica are so called because the females deposit their eggs in the nests, or in or on the larvae, or sometimes in the eggs, of other insects, thus enabling their own larvae to live as parasites upon those insects. The group also includes those parasites of plants, the gall-wasps, so called because they induce the victimised plant to produce characteristic swellings or galls. Of the insects which parasitise other insects the best known are the ichneumon-flies (which are, of course, not flies), the velvet ants (which are not ants), the cuckoo bees and the cuckoo wasps, and, by name at any rate, the chalcid wasps. The chalcid wasps are minute and very few of them are easily recognisable by the amateur. But some are: which is more than can be said for the cynipid wasps, which are the gall-wasps. These are equally minute and somehow much more anonymous. To expect anyone, other than a specialist in cynipid wasps, to recognise these tiny creatures apart, to distinguish between species and name them correctly, would be asking altogether too much. But some of their galls are familiar to us all.

The use of the words 'their galls' may, I suppose, be taken to imply that the wasps make the galls. It is important to understand that the wasp does not make the gall. It induces much, but it makes nothing. It is a gall-causer, not a gall-maker.

I suppose that I ought also to point out here that not all galls are caused by cynipid wasps. I have already mentioned that some saw-flies cause galls – the familiar red bean-shaped gall on willows is caused by a saw-fly – but it must not be thought that galls are caused only by members of the order Hymenoptera. That is very far from being the case. Many true flies cause galls: the gall-midges, in particular, are parasitic on a wide range of different plants. A number of aphids also cause galls on a wide variety of plants: the 'purse-gall' on poplars and the 'pseudocone-gall' on spruces are caused by aphids. Some plant lice, one or two of the moths, and a few beetles also cause galls. The galls on the roots of turnip and cabbage and other Brassicas are well known to gardeners, not all of whom know that they are caused by a beetle. Nor are insects the only gall-causers. The eelworms which cause galls on swedes, turnips, tomatoes and other cultivated plants are very troublesome pests. It is fortunate that there are not many species of gall-causing eelworms. The gall-causing mites, on the other hand, are very numerous and attack an astonishing variety of plants. By far the best known of the mite galls, because by far the most conspicuous, are the 'witches' brooms', which are most frequently to be seen on old birch trees. 'Witches' brooms' are not confined to birch. They also occur on the cherry. When they do they are caused by a fungus, not by a mite. It will be seen that galls and gall formation is a large and specialised subject outside the scope of this book.

Nevertheless, something must be said about the gall-causing Hymenoptera. And the first thing that must be said is that we do not fully understand how the galls are formed. And this is really very strange because galls have been known and studied from the earliest times: not merely because so many are attractive to look at, but also because in Grecian and Roman times they had a considerable commercial value, being used in dyeing and in the dressing of hides. The Greek philosopher Theophrastus (378–286 B.C.) alludes to the superior quality of Syrian 'gall-nuts' and the Roman historian Pliny the Elder (A.D. 23–79) also mentions them as being excellent for their purpose. The ancients believed that the plants made the galls themselves and that the creatures found inside them came into being as the result of spontaneous generation. And not only the ancients: that belief was commonly held in Britain until the latter half of the seventeenth century. John Gerard (1545–1612), the famous herbalist, supported the view held by farmers that 'oke-apples being broken asunder foreshow the sequell of the yeare' by the living things found in them. 'If they found an ant they foretell plenty of graine to ensue, if a white worm or maggott, murren of beastes and cattell; if a spider, then say they, we shall have a pestilence or such-like sicknesse amongst men.' It was not until the famous Italian anatomist Marcello Malpighi (1628–94), whose training did not dispose him to belief in spontaneous generation, announced, after years of close observation, that galls resulted where punctures were made in plants by certain kinds of insects that men ceased to credit them with supernatural attributes. Malpighi believed that the insects concerned injected a fluid into the tissue of the plant which caused it to swell in a particular way. He thought that the gall was a swelling similar to that which results after a human has been stung by a bee or a wasp. Unfortunately time has shown that all is not quite so simple as that.

But, at least, it is certain that the plant does not itself make the gall. If it were to do so, whether to seal off the wound by an insect or for any other reason, then we should expect galls to be specific to the plant. We would expect all galls on oaks to be alike, all galls on willows to be alike, all galls on roses to be alike, and so on. Instead we find a fantastic variety of galls occurring on the oak, which of all trees has the greatest number of different kinds of galls, and these galls are specific to the insects inducing their formation. The expert, looking at a gall, can tell by its form that it will contain the egg or the larva of a particular species of insect. He has no need to open up the gall to make sure. Indeed, there is an American willow which is parasitised by no fewer than ten different species of gall-flies. The flies themselves are so much alike that their separate identification is extremely difficult. Each, however, causes a very distinctive gall to arise and these galls provide an easy and certain means of identifying the flies.

Knowing that certain insects cause galls is one thing: knowing how they cause them quite another. There can be no doubt that the bean-gall on willows is caused by the drop of fluid injected into the tissue of the leaf by the saw-fly as she lays her egg. These galls develop very rapidly and are fully formed before the larva leaves the egg. The egg itself is not the cause of the gall forming, for it has been shown that if the egg is destroyed immediately after it is laid the gall will still continue to develop. The cause must, therefore, be the fluid. But this is about the only case of gall formation in which the cause is known virtually beyond doubt. In the case of the gall-wasps it would seem to be quite clear that the puncture of the leaf is not the cause and there is no evidence that, if a fluid is injected during egg-laying, it plays any part in gall-formation. It also seems to be beyond doubt that the presence of the egg in the plant tissue is not in itself sufficient to cause the gall to grow, because the gall does not begin to form until after the larva has left the egg. And it may be a very long time, in some cases months, between the laying of the egg and the hatching of the larva. It would seem obvious, therefore, that in the case of the gall-wasps it is the presence of the larva that induces the gall to form. It is thought probable that the larva does this by exuding a secretion. This is a reasonable supposition – it has not yet been proved beyond question – but it still leaves a great many questions unanswered.

It has been shown that if the saw-fly fails to lay her egg in the cambium layer no bean-gall will arise on the willow leaf. The egg simply fails to hatch. It does not seem to be known if it is as essential for the gall-wasp to lay her egg in the cambium layer: presumably not, since the gall does not form until after the egg has hatched. But, presumably, the larval secretion must affect the cambium and other actively dividing cells of the plant in such a way as to modify normal growth. Consider only three of the galls on oaks – the oak-apple, the marble-gall, and the spangle-gall (which appears on the underside of the leaves between July and September) – which are apparently induced by larvae. These three galls are so very different in formation that one would have thought it possible to detect and identify the larval secretion that is said to induce them. So far as I am aware, this has not been done. The cynipid wasps do not, of course, confine their attentions to oaks. The reddish marble-like galls on the stems of ground ivy are caused by a gall-wasp, another causes the well known 'robin's pincushion' or moss-gall on the dog rose and the sweet briar and their garden relatives, and two more cynipid wasps, virtually identical in appearance, cause the two pea-like galls, the one smooth and the othery spiny, which occur on rose leaves. You may well find these two galls on the same leaf, side by side and almost touching. It is difficult to believe that both can be caused in the same way at about the same time.

Many of the gall-wasps, and particularly those associated with the oak, exhibit what is known as alternation of generations. In each year there are two generations: one composed of females only, the other of individuals of both sexes. The bisexual generation is produced during the summer and gives rise to the unisexual generation which survives the winter. The extraordinary thing about this is that the females of the two generations are often so entirely different that they were originally considered to be totally different species of gall-wasps and were given different names by the experts. The galls of the two generations are also often very different in appearance and may arise on different parts of the same plant.

The story of the common oak-apple illustrates this phenomenon very well. If you will go into an oak wood towards the end of January and scrape away the earth near the foot of a tree, you will probably find (if you do not at the first tree, you certainly will at the second or third) on some of the rootlets you have exposed hard, brown, spherical growths, ranging in size up to half an inch in diameter. These galls may occur singly or in masses, when they look rather like bunches of grapes. If you cut one of the galls open, the wasp *Biorrhiza* will crawl out. It is always a female and always wingless. If you had not disturbed her, she would in due course have bored through the hard wall of the gall (and it really is very hard) and found her way up through the earth to reach the trunk of the tree. She would then have crawled up the trunk to the branches – a hazardous operation, for she is now exposed to all the vagaries of the weather and to attack by hungry birds – and then along the branches to the extremities of the shoots so that she can lay her eggs in the terminal buds. Dr Adler, who called gall-wasps 'gall-flies', even though he knew perfectly well that they were not, recorded his observation of Biorrhiza egg-laying thus: 'A fly was put upon a little oak, and soon began to prick a bud; when it had finished the first bud, it went on without interruption to another, and was altogether eighty-seven hours busily employed in laying its eggs. In these two buds I counted 582 eggs.' The buds are, indeed, sometimes attacked so fiercely that the tissues are destroyed and no gall appears. But if Biorrhiza does not allow her enthusiasm to run away with her the bud begins to swell early in May. Gall formation proceeds rapidly. The oak-apple appears in about a fortnight and has usually attained maturity by the end of the month. It then has a rosy colour and does resemble an apple in appearance. If you cut an oak-apple open, you will find that it contains a number of cells or chambers in each of which there is normally a small Biorrhiza larva. (There are usually also parasites present: but that is another subject which will be dealt with later on.) The wasps reach maturity in July and then gnaw their way through the gall wall to reach the outer wall. Their former presence is then revealed by their exit holes. The males are winged. The females are usually wingless, but may have rudimentary wings which are quite useless for flight. The female looks very much like her mother, but is a good deal smaller. After mating, which takes place at the end of July, the females crawl down the trunk, burrow into the soil to find the rootlets, which they pierce with their ovipositors and lay their eggs within. When the eggs hatch, the larvae give rise to the grape-like galls I have already described. The cycle begins all over again.

Excepting in minor details the life history of the oak-apple gall-wasp is the same as that of all gall-wasps in which regular alternation of generations occurs. In all, the larvae of the bisexual summer generation develop in soft galls of rapid growth, which are formed when the supply of sap is abundant. In all, the larvae of the parthenogenetic winter generation develop in hard galls of slow growth, which are formed in autumn when there is a diminished supply of sap and which provide the requisite shield for the helpless larvae through the harsh months.

Parthenogenesis means the development of young from unfertilised eggs. Just why it is that certain insects can produce young from unfertilised eggs, whereas all others and all other forms of animal life require the fusion of an egg and a sperm, is an unexplained mystery, though it is a well-demonstrated truth. The gall-causers among the Hymenoptera, when reproducing by parthenogenesis, nearly always give birth to females only. Indeed, in some of the saw-flies and in some of the gall-wasps the male has disappeared altogether – generation upon generation has been bred in entomological laboratories without a single male ever having been produced – and there are a number of species well on the way to achieving this evidently desirable end: *Rhodites*, the wasp which causes the robin's pincushion gall, for example, very rarely produces a male. On the other hand, the Gooseberry saw-fly produce only males by parthenogenesis. It is very easy to breed out the adults of this saw-fly, any amateur can do it, since the larvae are to be found on gooseberry bushes everywhere. If, when they have been bred out, steps are taken to keep the females separate from the males, they will lay virgin eggs and these will produce males in every case. If females and males are not kept apart, the females will lay fertilised eggs and these will produce females in every case.

So much, and it is not much, is known. There are yet mysteries to be solved. One concerns the familiar marble-gall on oaks. This is caused by the gall-wasp *Cynips kollari* and only females are known to emerge from marble-galls. Many thousands of these galls have been bred out in laboratories and nobody has yet seen a male. On the face of it, it would seem that Cynips has succeeded in abolishing the male. But, despite the thousands of laboratory experiments, it is by no means certain that this is so. The marble-gall is a naturalised British citizen. These galls were brought to Exmouth in 1834 for dyeing purposes in connection with the cloth-making industry. Previous attempts to introduce the smaller Aleppo gall, which occurs on Turkey oaks and which is the gall used by the Greeks and Romans for dyeing purposes, had failed because the wasp could not apparently survive in Britain. Cynips, on the other hand, took to the native oaks like a duck to water and in a very short time the marble-gall had spread all over Britain. It has been suggested that Cynips has not, in fact, succeeded in abolishing the male. Long ago it was suggested that the gall-wasp *Andricus*, which causes the very different and smaller galls on Turkey oaks, is the bisexual generation of Cynips. In 1897 Professor Beijerinck imprisoned a few Cynips (all parthenogenetic females remember) on saplings of Turkey oak and watched them oviposit in the buds. From these buds Andricus, male and female, emerged in due course. But Beijerinck failed to induce these Andricus to oviposit in buds of the common oak. Many other people have repeated this experiment and achieved precisely the same result. So we still do not know for certain whether Cynips has really abolished the male.

CHAPTER THREE

THE PARASITES

If we do not fully understand how the vast majority of galls are caused and if we do not, perhaps, know quite as much as we should about the lives of the gall-causers, we do know a great deal about what goes on inside a gall.

In the last chapter I said that if you cut an oak-apple open you will find that it contains a number of cells or chambers in each of which there is normally a small Biorrhiza larva. I added that there are usually parasites present. As a matter of fact, it would be a very exceptional oak-apple if each of its cells contained only a Biorrhiza larva. If you cut a marble-gall open you will again find a number of cells or chambers, but only one will contain a Cynips larva. When it is first formed the marble-gall is a single-chambered gall housing just one Cynips larva. The other chambers (and there are usually several) are made by other insects, which make use of the gall. They are uninvited guests, who do not pay for their keep, and they are known as 'inquilines'. And it is the same with the 'robin's pincushion'. A well-formed robin's pincushion may measure an inch or more in diameter by the end of July. Cut such a one open and you will find that it contains a number of cells, each harbouring a single larva of the wasp Rhodites which caused its formation. That, of course, is what one would expect. But as with the oak-apple and the marble-gall – as with all galls, no matter how small – there will be other insects as well: both inquilines and

Larvae of the Pine Sawfly (*Lophyrus pini*)

All too familiar to the gardener: larvae of the Gooseberry Sawfly

Female ichneumon *Rhyssa persuasoria*

Dead caterpillar of the Scarce Vapourer Moth surrounded by the fully fed larvae of the Apanteles that killed it

parasites. There is much, much more to a gall than meets the eye. Each gall, in fact, supports a whole association of different insects. Each is, indeed, a small world of its own: highly competitive and completely ruthless.

Let us have a closer look at a robin's pincushion. Rhodites causes the gall to form. Another cynipid wasp, which causes no gall formation of its own, waits until a gall has been formed and then lays its eggs in it. It behaves, in fact, very much in the manner of the cuckoo which, being unable to build a nest of its own, lays its eggs in the nests of other birds. (Our cuckoo did once build its own nest – there are a number of cuckoos in other parts of the world which still do – and it is a nice point whether its present inability to build is a sign of degeneration or the result of progressive thinking. The same point, of course, applies to the wasp.) Both Rhodites and its inquiline are subject to the attentions of a variety of insect parasites. One of the most beautiful of the chalcid wasps, *Torymus*, resplendent in metallic green and burnished copper, specialises in the actual gall-causer, using her long bristly ovipositor to pierce the gall to reach the Rhodites larvae in which she lays her eggs. Another chalcid, *Oligosthenus*, specialises in the larvae of the inquiline, using exactly the same method. Neither ever seems to make a mistake – Torymus never seems to lay her eggs in the inquiline larvae nor Oligosthenus in Rhodites larvae – and that when you come to think about it, is little short of miraculous. The larva of yet another chalcid has been noted making its way from cell to cell within the gall, killing and eating the larvae of the inquiline, but ignoring those of the rightful owner. (I suppose that, strictly speaking, this last chalcid should be described as a predator rather than a parasite.) It must not be thought that the chalcid wasps have matters all their own way. That is very far from being the case. At the outset they face competition from an ichneumon which specialises in the larvae of Rhodites, from another which specialises in the larvae of *Oligosthenus*, from a third which specialises in the predator larva. Nor is that all. The successful chalcid and ichneumon parasites are themselves parasitised by one or more species of hyperparasites, both chalcids and ichneumons. And these hyperparasites in their turn have their parasites.

> Great fleas have little fleas upon their backs to bite 'em,
> And little fleas have lesser fleas, and so on *ad infinitum*.

If all this sounds a bit complicated to you: well, it is. But do not think that it is in any way unusual. As I have said, every gall of whatever kind, no matter how small, is a world of its own and every one is every whit as ruthlessly, as savagely, competitive as that of the robin's pincushion.

If you read that paragraph again, and think about it, you may wonder how any of the Rhodites larvae ever manage to survive. The toll is enormous, of course. A. D. Imms has recorded that E. McCallan kept 1,015 robins' pincushion galls, which he had collected from a number of localities, on sand in glass jars covered with muslin, which he stored in an open summer-house. Only 815 of the galls produced insects, a fact which prompted Imms to comment that the conditions in which the galls were stored may not have been completely satisfactory. Be that as it may, the total number of insects that McCallan bred out was 24,393. Of these some 6,007 were Rhodites and the rest were either parasites or inquilines. In other words, approximately one in four of the insects bred out was a Rhodites: a survival rate more than sufficient to ensure the survival of the species. And it may be taken as certain that the survival rate among the other 18,000 insects would have been roughly in proportion.

If you should consider breeding out the inhabitants of galls, then I would not recommend you to start with the robin's pincushion, which, though one of the most rewarding in terms of different species, is, in my opinion, one of the most difficult. Many galls vanish with the leaves: the robin's pincushion is one of them. I do not know if McCallan's storage conditions, referred to above, were unsatisfactory or not. I should have thought not. I should myself have thought that he did remarkably well to secure such a large hatch, for I suspect that the real difficulty about breeding out robin's pincushion is not the method used for storing the galls but the fact that the leaf must be severed from the living plant. This, I would think, must have an adverse effect on the gall and its inhabitants. The ideal galls for breeding out are the oak-apple and the marble-gall. Of the two the marble-gall is, I think, much the better. The best time to collect them is in winter when they are most conspicuous on the oaks: any time from December onwards, but the later the better and late January to mid-February is, perhaps, ideal. Only those galls that are whole are worth collecting. If there is a small hole, Cynips will have left voluntarily. If there is a large hole, Cynips and its parasites and inquilines will have been removed by hungry birds. The large hole and the neat straight tunnel to the centre of the gall is the work of woodpeckers and nuthatches. You will usually find a few galls, one side bashed away, on the ground at the foot of the tree. This rather clumsy work is that of tits. As to storage: I think that the best method is to take the twig bearing the galls, not to separate galls from twig, and to place it upright in soil or sand in any sort of container, fitting a muslin cover. Where you place the container does not, I think, matter in the least, so long as it is sheltered from the rain. I believe in removing the cover once a week and giving the galls a misting with a scent spray.

The chalcid wasps do not, of course, confine their attentions to cynipid wasps in galls. Some specialise in the caterpillars of butterflies and moths, some in the pupae of butterflies and moths, while some are so minute that they are able to lay their eggs inside the eggs of other insects. The degree of specialisation is considerable. There is, for example, one chalcid which specialises in the newly changed chrysalis of garden white butterflies. A French entomologist, F. Picard, was the first to record how the wasp will settle beside a caterpillar which is about to pupate and wait, perhaps for several hours, until it does so. As soon as it does she mounts the chrysalis, which is still soft and penetrable, drives her ovipositor into it and lays a number of eggs. She is interested only in the caterpillar about to pupate: all others are ignored. The smallest of all chalcid wasps – they are commonly called fairy-flies (though they are so small that most people never notice them at all) – are less than one-fiftieth of an inch long. They specialise in laying their eggs in the eggs of plant lice, which are themselves so small that it is difficult to believe that any parasite could develop within one. Nevertheless these do, and greatly benefit the gardener by so doing. Another chalcid specialises in the eggs of the common Silver Y moth, another in the eggs of the Small Ermine moth, and so on. In every case the eggs of other species are wholly ignored: all these wasps are dedicated specialists. And they take every precaution to see to it that the egg in which they have laid their own egg is left alone by other females of their species. Before laying, the chalcid will crawl all over the egg, pressing her abdomen down upon it, evidently leaving her scent. Only when she is satisfied that she has made her presence abundantly obvious will she plunge in her ovipositor and lay her egg. The next chalcid that comes along will alight on the egg and almost immediately take off again, well aware that it will not make a satisfactory home for her young. I should doubt if two individuals ever parasitise the same egg, but the Silver Y specialist usually lays several of its eggs within a single host-egg.

In the last chapter we discussed the phenomenon of parthenogenesis. Many chalcid wasps, perhaps all of them (I do not know), are parthenogenetic, laying virgin eggs from time to time. These virgin eggs always develop into males. If mating occurs and the eggs are fertilised, then the progeny always consists solely of females. In addition to being parthenogenetic, however, the chalcid wasps have developed another, even more astonishing, method of reproduction: the phenomenon of polyembryony.

Let us take the Silver Y moth as an example. The chalcid wasp lays its egg or several of its eggs inside the moth egg. When the moth egg hatches the caterpillar that emerges carries within it one or more eggs of the chalcid wasp. These chalcid eggs, instead of developing directly into larvae, as one would expect them to do, develop (each one separately) into a peculiar sausage-shaped mass of embryonic cells. Each mass of embryonic cells continues to grow and divides to form a chain of separate embryos. During the dividing of the embryos they are fed by nourishment absorbed from the tissues and body fluids of the host. When a dozen or more embryos (the number depends upon the species of chalcid) have been formed, the mass of embryonic cells ceases to divide and each embryo develops into a parasitic larva. Thus from each egg laid by the parent parasite within the moth egg a dozen or more, usually many more, young may develop. The maximum number known to originate from the division of a *single* egg is a thousand! This occurs – it does not always occur, of course, but it is probably not nearly so uncommon as one might imagine – in the chalcid *Litomastix*, which is the one that specialises in the eggs of the common Silver Y moth. (Not all chalcids are as productive as this, of course. The wasp which specialises in the eggs of the Small Ermine moth, for example, lays eggs which give rise to 80–100 larvae each: by comparison a modest achievement, but still one bordering upon the miraculous.) If several eggs are laid within a single host-egg – as frequently happens in the case of Litomastix, but not, apparently, in the case of the chalcid parasitising the Small Ermine moth – then really large numbers may be reared within the body of one caterpillar. More than 3,000 adult Litomastix have actually been counted after they had emerged from the body of one dead Silver Y caterpillar. 'Dead' does not really give the right impression, for there is no body: the caterpillar is by then no more than a shell. It died long before: when the parasitic larvae pupated; and they did so when they had destroyed the body, leaving only the outer shell to afford them some shelter.

All this, remember, happens within the body of the unfortunate caterpillar. Parasites which live inside the bodies of their hosts are known as endoparasites. (Parasites which, like fleas and lice, live outside the bodies of their hosts are known as extoparasites.) The endoparasites of insects generally live in the blood-filled spaces making up the body cavities of their hosts. They wander about in these blood-filled spaces avoiding the vital organs and feeding on the blood or on the less vital tissues, such as fat. As a result of this skilful behaviour on the part of the endoparasites, the host is not killed outright, but continues to feed and to grow, though usually more slowly than one not so burdened. The parasitic larvae are thus ensured of a regular food supply throughout their developmental period. On the completion of their development, but before they pupate, the larvae of many endoparasites burrow outwards through the body-walls of their hosts, killing them as they do so. The larvae of the chalcid wasps are exceptions to this final feast before they do so. Now, it is easy enough to understand how these endoparasites feed. But how do they breathe? Within the bodies of their hosts they are sealed off from the fresh air, but, if they are to live, they must breathe somehow. As they live in the blood-filled spaces of the host's body cavities, they are, in effect, leading a semi-aquatic existence and so they naturally show many of the respiratory adaptations developed by genuinely aquatic larval. They breathe through the skin. Oxygen simply diffuses from the blood of the host into the blood of the parasite. This is amply sufficient for the needs of the parasitic larvae, which do not have to expend much energy moving about within a closely confined space.

The marked degree of specialisation, the rigid conservatism, exhibited by the chalcid wasps occurs throughout the parasitic members of the Hymenoptera; with the possible exception of the velvet ants.

The velvet ants – there are two species in Britain – are wasps. In both species the female is completely wingless and can easily be mistaken for an ant by a casual observer.

(If you are ever in doubt as to whether an insect is an ant or not, look at the hind-body. In all ants this consists of a very slender waist, known as a 'petiole', which joins it to the thorax and continues backwards into a globular gaster. The petiole bears an enlargement, a hump, which produces two waists, one before and one behind the hump. This feature, which gives great mobility to the hinder part of the abdomen, is not found in any other insects with petioles. In some species of ants, by the way, there are two such humps and, therefore, three waists.) Remarkably little is known about the habits and life history of velvet ants. All the books of reference seem to be agreed that the velvet ants are parasitic on bumblebees, the female laying her eggs in the bodies of the bee larvae and herself feeding off the store of honey. John B. Free and Colin G. Butler in their *Bumblebees* in The New Naturalist Library devote a chapter to 'the more common and interesting predators, parasites and scavengers of bumblebees' nests'. They do not mention velvet ants at all, which, in view of the insistence of the reference books, is surely rather surprising. Obviously, these two great authorities do not regard velvet ants as serious parasites of bumblebees. I have myself no knowledge of velvet ants in the nests of bumblebees – which may well mean no more than that I have not looked closely enough – but I have more than once seen them entering or leaving the nests of digger wasps in my garden. Maybe, the velvet ants are exceptions to the general rule of parasitic specialisation.

The bumblebees are parasitised, of course. The real specialists are cuckoo bees, which, in general appearance, are very like true bumblebees. Indeed, it is not possible to distinguish the males from male bumblebees at sight and you have to be pretty expert to distinguish the females, which generally resemble the bumblebees they parasitise in shape and colour, though they are a little less hairy and have the abdomen rather more curved. But the feature which most surely distinguishes the cuckoo-bees of the genus *Psithyrus* from genuine bumblebees is the lack of pollen-gathering equipment: the hind shin is not shaped like a scoop and is hairless. Still, if you are just looking at a bee in the garden, you need pretty sharp eyesight to notice a little thing like that. A much better guide, I think, is behaviour. The bumblebees are always busy, working hard to support the colony. There is an air of urgency about everything they do. Psithyrus has no work to do. She flits from flower to flower in the most leisurely manner. The air of urgency is wholly absent.

Psithyrus has no work to do because, in true cuckoo fashion, she gets it all done for her. Only the females survive the winter and they emerge from hibernation later than the female bumblebees. Indeed, Psithyrus does not put in an appearance until the bumblebee nest is equipped with workers. Then she enters to lay her eggs and take up residence. There are records of fights before she succeeds, and even a few records of failure to achieve entry (the rightful owners killing the would-be invader), but usually she seems to get in without trouble and, once in, she is at home and allowed to come and go as she pleases. Her family are raised by the bumblebee workers (she produces no workers herself) and she and her young feed from the bumblebee honey-pots. No bee, one would think, could ask for more. But there seems to be little doubt that after a few weeks she murders the rightful owner, stinging her to death, so that she and her brood may receive the undivided attentions of the workers.

Our other cuckoo-bees are Coelioxys, Melecta and Nomada. These specialise in the solitary bees which we will be considering in the next chapter. Melecta is a grey bee, resembling the potter bees which she parasitises quite closely, though she is rather larger. Potter bees are plentiful, so Melecta has every opportunity for practising her art: but she seems to be a bit of a failure, because she is really quite uncommon. Coelioxys specialises in leaf-cutter bees, resembles them closely, and is a pretty successful parasite. Nomada, specialising in the mining bees, makes no attempt to look like them, but is also pretty successful. All the Nomada are hairless (the mining bees are noticeably hairy) with clearly defined waists and sharply defined yellow and black stripes on the body. They look like wasps, not bees: and it may be that this is the explanation of their success. All three lay their eggs in the cells of their hosts, but make no attempt to interfere directly with the cell-maker. Indeed, Coelioxys and Melecta are at great pains to slip in and out unseen. The larvae of all three begin by eating the food provided for its young by the cell-maker and end by eating the young as well.

And this is also the way of the cuckoo-wasps, which are popularly known as ruby-tailed wasps. None of these bothers about resembling its host or about concealment. None is black and yellow in the traditional wasp manner. The most common colour is red and black or more red than black, but whatever the colour – and there are some gorgeous blues and greens and coppers, but always with some red or black somewhere – they are always bright and metallic. The ruby-tailed wasps are specialists in the solitaries, bees and wasps, never interfering with the adults, but always trying to slip in and out while the owner is absent. Again, their larvae in due course devour the larvae of the host.

But the greatest specialist of all is, surely, *Rhyssa*, the largest of the ichneumons, whose particular prey is the giant wood-wasp. As we have seen, the Sirex larva spends its life boring its way through conifer wood. Rhyssa has an ovipositor $1\frac{1}{2}$ ins long – longer than her head and body which together measure but $1\frac{1}{4}$ ins – with which she bores through the bark and the solid wood to reach her victim in its tunnel. Her egg is actually laid on the Sirex. Now this, you must admit, is a very remarkable achievement.

How does Rhyssa detect her victim hidden deep in the wood? It seems to be generally accepted that she does so by her sense of smell, that she detects through her antennae the scent of the Sirex beneath her. The process has been watched many times. Rhyssa explores the surface of the trunk very thoroughly with the tips of her antennae. When she locates a Sirex she becomes unmistakably excited, jigs about for a moment or two, seems to stroke the wood with the tips of her antennae, and then settles down to bore. Scent certainly would seem to play a part, but it surely cannot provide the whole answer. If a Sirex larva smells so strongly, then it must surely leave its scent in the tunnel and for some time. If Rhyssa worked on scent alone she would surely often bore through to a tunnel recently vacated by a Sirex. Perhaps she does, but there is no evidence that she does. Working on scent alone, she would surely often bore through just ahead or just behind her victim. Perhaps she does, but again, there is no evidence that she does. There is, on the other hand, plenty of evidence that she hits the target time after time. Can it be that she gets a sort of 'geiger-counter' response through her antennae?

Be that as it may – and there is much here that we do not properly understand – there can be no doubt that she lays her egg on the Sirex larva much more often than not. When the egg hatches the larva attaches itself to the Sirex larva by means of its mouth-parts, which perforate the skin, and then lives by sucking its host's blood. The Sirex larva meanwhile continues to feed and to tunnel, but sooner or later it becomes exhausted and dies of starvation. The Rhyssa larva pupates and becomes adult within the wood-wasp tunnel, and the adult then proceeds to gnaw its way out of the tree trunk. Nobody knows how long all this takes. Nobody knows how long the Rhyssa eggs, which are laid in June and July, take to hatch. Nobody knows for how long the Rhyssa larva must feed before it is ready to pupate. It is a safe bet that Rhyssa eggs hatch very quickly, but the future of the larva must surely depend to a great extent, if not entirely, upon the age of the Sirex larva at the time of the attack. A very young Sirex, for example, would not be able to support a parasite: one about to pupate might not allow the Rhyssa time enough to mature. And, of course, at the end of the road there can be no guarantee that the adult Rhyssa will be close enough to the surface to be able to gnaw its way out before it dies from exhaustion. Presumably, for these and other reasons – such as the egg being laid on a Sirex larva already parasitised by another Rhyssa (in which case all three larvae will die) – there must be many failures. But the undoubted success of Rhyssa as a species is evidence enough that the whole business is not nearly so haphazard as it would appear to be.

The word 'parasite' has an unpleasant connotation for humans. We think of the parasite as a lazy creature leading a degrading life. But the literal meaning of the word is 'one who eats at the table of another'. The parasitic insects take immense trouble to find a suitable table and to place their eggs on or in it. Considered from their point of view, this is a most praiseworthy action: they are attempting to give their young a better, a safer, start in life. Among the Hymenoptera the evolution of parasitism proved to be a great step forward: the first step towards personal involvement with the young.

CHAPTER FOUR

THE SOLITARIES

The remaining group of the Apocrita is the Aculeata: the group of stinging insects. This consists of the ants and the bees and certain wasps in which the ovipositor has been modified into a sting or defensive organ and is no longer used for egg-laying, the eggs being ejected from the body of the female through an opening at the base of the ovipositor. A special characteristic of the Aculeata group is the care and attention which the majority of its members give to their young: a characteristic which in some of the bees and wasps, and in all the ants, has given rise to an elaborate form of social life.

As I have said, it is my opinion that the first step towards personal parental care for the young among the Hymenoptera was the evolution of parasitism. (I must stress 'among the Hymenoptera' because true parental care for the young has also evolved in one other, wholly unrelated and much more primitive, order of insects, the Isoptera or Termites, commonly known as 'white ants'.) There is, of course, no direct connection between parasitism and a social life. Many people, indeed, would deny that there is any connection at all. But no one, I imagine, would deny that there is a connection between a nest and a social or community life; for obviously there can be no family life, let alone a community life, without a base within which or from which the members can operate. If you accept that there is such a connection, then you may find it significant (as I do) that, although parasitism has arisen in three orders of insects – Coleoptera (beetles), Diptera (flies) and Hymenoptera: I am referring to parasitism of insects by insects, of course, not to parasitism of mammals or birds by such insects as fleas and lice, which is quite another matter – it is only in the Hymenoptera that we find deliberate nest-building and it is only in the Hymenoptera that we find organised social or community life.

The most primitive insects simply scatter their eggs in the particular environment in which they live: they are not especially concerned that they should fall on the table and they do not bother about protection at all. Many more are at great pains to place their eggs on the table, on some part of the environment which forms the normal food of their larvae. Butterflies, for example, are most careful to lay their eggs on the leaves of the food plant of their caterpillars. Peacock and Red Admiral, Comma and Small Tortoiseshell feast at buddleia and sedum and Michaelmas daisy. They are careful, however, to lay their eggs on the leaves of the humble stinging-nettle, a plant which they ignore at all other times, but which is the food plant of their caterpillars. This is a considerable advance on the indiscriminate scattering of eggs in the immediate neighbourhood of the food plant. But, though these butterflies see to it that the correct food is readily available for their caterpillars, they do nothing to protect either their eggs or their caterpillars from danger. The Hymenoptera early took a further step. The saw-flies, the wood-wasps and the cynipid wasps do not lay their eggs on the table, as do the butterflies, they place them in it. The saw-fly, making its slit in the leaf and placing its egg therein, is providing that egg with a protective covering which the egg of the butterfly lacks. The fact that this action causes a gall to form may be fortuitous – we do not know that the saw-fly and the cynipid wasp know, when they puncture a leaf or a terminal bud, that they are going to cause a gall to form as we do know that the common wasp intends to build a nest – but the fact that a gall is formed is an additional protection for the egg or for the larva. Of course, there are plenty of other insects eager to take advantage of the gall. The vegetarians, unable to cause a gall to form themselves, soon learnt to make themselves at home in the galls of others and, of course, their predators followed them. But all that is beside the point. The point is that the primitive Hymenoptera took steps to afford some protection for their eggs and their young. The gall, whether fortuitous or not, is the first step to the specially prepared, deliberately constructed, nest.

The earliest deliberately constructed nests – there can be no doubt that they are very much earlier than the colonial nests – are those of the solitary bees and wasps. With the construction of these nests personal involvement with the young is greatly increased for both the bees and the wasps, for not only do they build a nest to afford protection for the young, they also provision it for the sustenance of the young during the larval period. The solitary bees concerned are the potter bees, the leaf-cutter bees and the mining bees: the solitary wasps are the potter and mason wasps, the sand wasps and the digger wasps. Both groups work in much the same way, but the bees are all vegetarians while all the wasps are carnivorous.

The word 'solitary' must not be taken too literally. Many of these so-called solitary bees and wasps live in colonies – indeed, hundreds of individuals may sometimes be found in one small area – but each individual does its own work regardless of its neighbours. Each female builds her own nest and is quick to resent interference by others. In fact, each individual attends strictly to its own business and has nothing whatever to do with its neighbours. There is no communal life. When colonies occur, they do so only because conditions at that spot are particularly favourable for nesting.

The solitary bees and most of the solitary wasps are nesters in holes. Many nest in holes in the ground which they bore for themselves. Naturally, there is a marked preference for light soils with good drainage, and sandy districts are particularly favoured. But many more, especially among the bees, are all for saving themselves trouble whenever they can and eagerly take advantage of beetle-holes and nail-holes in beams and fences, of key-holes, of crannies in walls, and similar situations. This habit, fortunately, makes some species very easy to encourage and easy to watch.

Let us consider the bees first. The potter bee (so-called because of the shape of her cell) digs herself a burrow in a sandy bank, if there is not a ready-made hole handy, and then excavates a small chamber from its side. In this she builds a small pot, which she lines with a paper-like substance and then fills with a mixture of honey and pollen. On the surface of this mixture she lays a single egg, closes the mouth of the pot, and then sets about excavating another chamber and building another pot. When her last egg has been laid and the last pot sealed, she dies. She never sees her young. The leaf-cutter bees behave in the same way. Though their cells differ in structure, they too are provisioned with a mixture of honey and pollen. Again, the bee dies when she has laid her last egg and never sees the result of her labours. The leaf-cutter bees, as a whole, much prefer a ready-made hole and will bore one for themselves only as a last resort. The mining bees, on the other hand, are all confirmed diggers in soil. Some are most industrious diggers and will sink shafts to a depth of 10 inches or so. For this sort of work the right soil is absolutely essential and so you will find the same patch of ground being used year after year: sometimes by very large numbers of bees. Again, most mining bees die when the last egg has been laid and never see their young. But in one genus, *Halictus*, this is not always the case. In one British species, *Halictus calceatus*, and there are others elsewhere, there are two generations in a year and the fertilised females of the second (autumn) generation survive the winter to become the founders of the summer generation of the following year. The great importance of this will be explained later on.

With the exception of the potter wasps (so-called for precisely the same reason as the potter bees), all the solitary wasps are nesters in holes. The vast majority prefer to make their own holes either in the ground or in decaying wood, but a few extract the pith from the stems of plants and build their cells in the tubes thus fashioned. The potter wasps, however, make their beautiful vase-shaped cells of mud or sand and tiny pebbles, which they mould together with their saliva and then attach to the stems of woody plants, especially heather, or to wooden posts or beams.

All the solitary wasps provision their cells as do the bees. But the wasps are carnivorous and feed their larvae upon whole insects. This has led to a quite remarkable degree of specialisation. Some species specialise in spiders, some in caterpillars, some in weevils, some in flies of various sorts, some in bees, some in aphids, some in the nymphs of the frog-hopper dragged from the refuge of their 'cuckoo-spit'. And by specialise I really do mean specialise. A spider-hunting wasp will not give a caterpillar a moment's consideration; and so on. The reason for this is, I think, that these solitary wasps provision their cells with whole insects, which are alive and not dismembered in any way. But the prey is stung before it is placed in the cell. The sting paralyses it and renders it incapable of activity. (But not of all movement: an occasional twitch is permitted, and it seems that this may be essential to encourage the larva to feed.) Now it may well be, indeed it is probable, that venom which is sufficient to paralyse, but not to kill, a spider will either have no effect or will have too much effect on a caterpillar or a weevil or a fly or whatever: and vice versa. What is required is that the prey be paralysed, not killed. The strength of the venom dictates the choice of prey.

Let us now consider two examples of solitary wasps in rather more detail. The potter wasp Eumenis is a caterpillar specialist. She builds her beautiful little pot, attaches it to a stem of heather, and then goes off to find a caterpillar. She specialises in little green caterpillars. When she has caught and paralysed enough to stock the cell – this usually requires about a dozen, for they have to last for the whole of the larval stage of her young and that is a period of months – she lays a single egg, suspending it from the ceiling of the cell by a silken thread so that it hangs just above the store of caterpillars, closes the door, and starts to build another pot. When the larva hatches it eats its first caterpillar while still attached to its egg-shell and suspended above the store. But it soon becomes strong enough to sever its connection with its egg-shell and to drop on to the store of caterpillars, where it remains until it has eaten them all. It then pupates and in due course the adult wasp gnaws its way out of the pot, mates, and then sets about building pots of its own. As with the solitary bees, Eumenis never sees her young. When her last egg has been laid and her last pot sealed, she dies.

Eumenis is the only wasp to suspend her egg from the ceiling of the cell (all the others lay their eggs directly on to the paralysed prey), but she is far from being the only wasp to capture and paralyse caterpillars. There are a great many caterpillar hunters, belonging to many different species of different families. Among them there is a wide range of specialisation. Eumenis goes in for small green caterpillars: they have to be small and they have to be green. Another specialises in caterpillars which feed only by night, spending the day hidden at the foot of their food-plants: but not so well hidden as to defeat the wasp, which appears to find them by scent. Another specialises in large caterpillars (so large that it has great difficulty in dragging them along), since it requires but one to stock each cell. And so we could go on: to each its particular speciality.

Anoplius is a sand wasp, which specialises in spiders. There are forty species of wasps in Britain which specialise in spiders. I have chosen the Anoplius because it is our largest, and probably also our commonest, species. It is a wasp with unusually long legs and it spends its time running very quickly indeed – it really is astonishingly fast – in short bursts over the ground, with brief flights between the bursts. It rarely flies more than a few inches at a time and never reaches a greater height than six inches or so above the ground. Any sort of spider, except apparently the web-spinners, is taken. (There are wasps which confine their attentions to the web-spinning spiders.) Now, a spider is a pretty formidable animal, armed with a lethal venom of its own; an animal well able to look after itself in a struggle with any creature of comparable size. Yet, so far as is known, no spider makes any attempt to bite Anoplius. All of them try to escape either by running away or by shamming death. The latter course is sometimes, but not often, successful. But there are few spiders fast enough to run away from Anoplius. The chase ends with the wasp getting a hold of the spider's back and then curling her abdomen round so as to sting it underneath. The wasp venom is potent and acts promptly. The spider is immobilised almost instantly. Only then does Anoplius think about making a nest. But first she hides the spider – usually a little way off the ground to protect it from ants – and then goes off to find a suitable place to dig a hole. During the digging she goes back to the spider several times, just to make sure that it is still there. When the hole is prepared she drags the spider to it, backs in, pulls it in after her, and then lays a single egg on its abdomen. Then she comes out and carefully closes the entrance to the burrow, smoothing the sand down until there is no sign that a wasp has been at work there. This performance is repeated every time an egg is laid, and it is thought that each Anoplius lays about twenty eggs. When her last egg has been laid and her last burrow securely closed, she dies. She never sees her young.

This applies to all our solitary wasps, save one. Ammophila, a sand wasp which specialises in caterpillars, is the exception. Ammophila digs a burrow in sandy soil and then goes off to find and paralyse a caterpillar. She drags this to her burrow, pulls it in after her, and then lays a single egg on it. She then closes the entrance to the burrow and goes off to prepare another one. So far, her behaviour appears to be typical. But it is not, for Ammophila does not close her burrows once and for all. She returns to them at intervals, opens them up, and pops in another caterpillar. She is the only British solitary wasp to see her young; the only one to continue care of the young. And this is a rather greater achievement than may appear at first sight. Ammophila does not make all her holes close together. They may not be very far apart (to our way of thinking), but they are not next door to one another. Ammophila has not only to remember where they are, she has also to know in what state they are; how much food is likely to be left in each, when each will be in need of replenishment. It is really a very considerable feat of memory. Ammophila works hard throughout the larval stage of her young. She dies only when they pupate.

Some way back I said that the preference shown by some solitary bees and wasps for ready-made holes makes them easy to encourage and easy to watch. This is done by providing artificial nesting-sites in suitable situations. Many years ago, sometime in the early 'twenties, I attended a lecture given by Professor Balfour-Browne on the habits of solitary bees and wasps. In it he described how he had made a 'bee-wall' in his garden and so had been able to watch the work of certain bees and wasps at close quarters. To begin with he had used seed-boxes filled with soil in which ready-made burrows were fashioned by inserting stems of elder from which the pith had been removed. Later he used air-bricks with glass tubes to fit the holes. I can remember to this day how excited I was by that lecture. I went straight home and made myself a bee-wall. And I have had one in every garden I have had ever since.

I use seed-boxes and air-bricks, for both have their advantages. Fill your seed-box with wet clay mixed with chopped straw – the straw prevents the clay from drying too hard – and then bore a number of holes about $\frac{3}{4}$ inch in diameter into this mixture so that they reach the bottom of the box. You will then have a number of burrows about three inches long. Into these burrows you can fit stems of elder from which you have removed the pith – elder is much the best plant to use because it is not difficult to remove the pith – and then fix the box vertically to a fence or a wall facing south, so that it gets as much sunlight as possible, because all bees and wasps are sun-lovers and do little or no work in the absence of sunshine. You will find that some bees and wasps will take over the unlined burrows, that others – and this applies particularly to some wasps – will only use the burrows lined with elder. I think that there can be no doubt that you get more different species at seed-box burrows – this is certainly true of wasps – than at air-brick burrows, but they are less easy to watch.

The air-brick has the great advantage that you can line the holes with glass tubes. If you get the right fit, then you can pull out the tube and have a look at what is going on inside. And you need not worry about whether the 'bee-wall' is going to be occupied or not. Fix it up a couple of feet from the ground, facing south, and you will have insects inspecting it within an hour or so. You may even have it occupied within an hour or so. And as to occupation: I have an air-brick with five rows of eight holes – you can get these bricks with different numbers of holes, of course, and it is a good idea to do so – and one summer I had thirty-eight of the holes occupied at the same time. Of the bees I have had Osmia, Megachile, Anthidium, and Chelostoma at my wall regularly year after year, and, of course, also Nomada the cuckoo-bee. All these bees are common where I live, but perhaps the most common, and certainly the most delightful, is Osmia. This is the Red Osmia (*Osmia rufa*). Its head is black, but the rest of its body is covered with a thick coat of reddish hairs. It looks very like a small bumble-bee, both in shape and hairiness. It is common on garden flowers from early April to June, and in those months you also often see it crawling on the ground, particularly in places where it has recently been dug or forked.

Osmia takes a very long while to inspect a bee-wall. She goes into hole after hole, disappearing altogether, and then, after what seems ages, backing out again to go immediately into another. Often she will go into the same hole or the same two or three holes time and time again. Having at last found a hole to her liking – or perhaps I should say, a tube to her liking; for the holes are surely all alike, whereas the glass tubes, being wrapped in brown paper (to protect them from the brickwork) may not be – she sits in the entrance for some considerable time; presumably to leave her body scent there. After that she comes out and flies backwards and forwards in front of it and facing it, quite obviously learning to recognise its position. This flight begins with a most careful inspection of the entrance itself, and then extends for some distance on either side and above and below. Such a flight may last for thirty seconds, after which she will re-enter the hole and disappear. Shortly she will reappear and the whole inspection flight will be taken again; and this may be repeated three or four times before she is satisfied that she knows her entrance. There will be other bees doing the same thing all around her, and quite likely at precisely the same time. But it is very rare for a bee to make a mistake and enter the wrong hole.

When a hole has been finally chosen Osmia sets to work

at once, bringing home a pellet of earth. She enters the tube head first and crawls to the end, puts the pellet on the floor and then works it with her jaws to spread it over the end wall of the tube. (If you have a tube without an end, then you must plug the gap with paper, but the modern phials in which you get pills save a deal of trouble.) She then fetches another pellet and another and another until she has built a wall of earth at the end of the tube. When this wall has been built, she goes off to sip nectar from the flowers – the nectar is converted into honey in her crop – and, while doing so, she collects masses of pollen which clings to the long hairs on the underside of her body. Then she returns to her tube, enters head first, crawls up to the wall she has built at the end, disgorges the honey from her crop, and backs out. At the entrance she turns round and backs in and all the way to the wall again, where she removes the pollen from her body by means of the brushes on her hind legs. Then she goes off for another load. When she has enough, she mixes honey and pollen into a paste, disgorges a final drop of honey on to the very centre of the paste, and then lays an egg on that spot of honey. Then she builds another wall to close off that cell, and starts all over again; building cell after cell until the tube is filled. When it is the tube is finally closed with a wall of earth about twice as thick as those forming the walls of cells.

All this you can watch right from the very beginning. All you have to do is to wait until the bee is in the tube and then withdraw the tube from the brick. This is easy to do. You will find that the brown paper wrapping sticks to the brick and not to the glass, so you get a perfectly clear view of all that is going on. And Osmia does not object. The fact that light suddenly floods into her burrow does not disturb her in the least: she just goes on working. And you need have no fear of being stung. Osmia is a most friendly and placid creature.

This is much more than can be said for the wasps which make use of my air-brick. Two members of the genus Odynerus do so: *O. callosus* and *O. parietum*. Both are common. In my experience *callosus* will have nothing to do with glass tubes, but she will take over an elder stem from which the pith has been removed. From the observation point of view this is not helpful, because you cannot see what is going on inside an elder stem. On the other hand, *parietum* takes kindly to a glass tube, but has the strongest objection to the tube being handled while she is inside it. She panics and flies off. Put the tube back and she will be back inside within a minute or so. I do not know how much interference she would put up with because I have never tried to find out. That sort of experiment does not appeal to me. All I can say is that I have not known *parietum* to desert a tube once she has started to build cells.

When a tube has been filled, it is a good idea to replace it with a fresh one. You can store filled tubes in an outhouse or garage, and so keep an eye on what is happening inside them. In the case of Osmia what happens is this. The egg hatches in ten days. The larva grows very rapidly, casting its skin at least four times during its life. At four to six weeks it is full grown and has exhausted the food supply in the cell. It now spins a silken cocoon which is reddish-brown to start with, but changes to white after a couple of weeks or so. This white stage remains unchanged until September when it gradually darkens and finally becomes black. Early in October the thin pupal skin splits and the adult bee emerges from it, but remains within the cocoon. It sleeps there through the winter, only awakening in March when it bites its way out of the cocoon and into the cell. She still has to escape not only from the cell but also from the tube. It will have been realised that the cells in these tubes are built one after the other, the first being that furthest from the entrance. One might imagine that the bee in this first cell would have a lot of trouble escaping from the tube. In fact, this is not so. Though the eggs may hatch at different times, the adult bees all wake up at pretty much the same time in the spring and escape from the tube in order one after the other.

If you have not got a bee-wall in your garden, then I hope you will get one as soon as possible. It will be a source of endless entertainment.

CHAPTER FIVE

THE SOCIAL BEES AND WASPS

The vast majority of insects are 'solitary'. As we have seen, in the solitary insects individuals take no interest in one another except during the brief period of mating. Each individual lives for itself and, except when thrown accidentally into association with some other animal, lives by itself. It does not even concern itself with the welfare of its own young. In fact, it rarely sees its own young: when it does, it fails to recognise them. It is an enormous step from that state of individual isolation to the state of a social being living in a community. How is it then that some insects have become 'social' and live in communities?

Solitary insects are not always solitary in habit. They sometimes gather together in large numbers. They are then usually described as 'gregarious'. It might well be thought that this gregariousness is the beginning of true social behaviour. In connection with insects, however, the term is often too loosely used. The dictionary definition of gregarious is 'living in flocks, given to association with others of the same species; fond of company.' The fact that solitary insects sometimes gather together in large numbers is not to say that these insects are gregarious. They do not normally live in flocks and there is absolutely no evidence to indicate that individuals have any liking for each other's company. Usually they are brought together by some local and temporary condition: moths by a street lamp at night, for example, or flies by a dead body. Mere crowding together is not necessarily a sign of gregariousness.

But there are gregarious insects: or, at any rate, insects which are gregarious for some part of their lives. The caterpillars of some butterflies and moths spin a silken shelter within which they feed together as a family. The caterpillars of some moths spin a smaller, stronger shelter which forms a more or less permanent home, from which they go out to feed and to which they return to rest. These are examples of genuinely gregarious behaviour. In each case, however, as the caterpillars become full grown they lose all interest in each other and disperse to pupate.

Another type of gregariousness affects certain insects periodically and arises from overcrowding. Locusts, for example, have two phases or forms: the one non-migratory and solitary, the other migratory and gregarious. There can be no doubt that the second phase, which gives rise to swarms numbering many millions which devastate the countryside over which they pass, is caused by crowding, by a greatly increased density in the population of the solitary form. A similar, but usually less spectacular, phenomenon occurs in caterpillars which, from time to time, devastate crops in different parts of the world. Again, there is a greatly increased density of population and the crop upon which this population feeds is rapidly destroyed. In such circumstances one would expect the caterpillars to scatter in all directions in search of food. Instead they stick together and the whole horde moves as one, as a disciplined army, in one direction to devastate a neighbouring crop, and then another and another, until finally the individuals disperse to pupate. These, locusts and caterpillars alike, are genuinely gregarious insects. But, even though they show signs of a rudimentary organisation, they are not social insects.

It has been said that gregariousness is the basis of social life: and this is undoubtedly true. But gregariousness and the development of social life are not necessarily closely related. Indeed, the one does not appear to lead directly to the other. There are many cases of gregariousness without the least suggestion of true social behaviour and there are many cases of social behaviour which seem to have developed quite independently of ordinary gregariousness. Certainly, gregariousness has never led to social life in insects.

Obviously, a social insect must be a gregarious insect: but that alone is not sufficient. The gregarious insect society knows not its parents, is no more than a more or less temporary collection of brothers and sisters. The essential step towards a social insect society is that the female should continue to have contact with her eggs after she has laid them and then with her young after they have hatched. As we have seen a few solitary bees and wasps have some brief contact with their eggs and with their young, but not such as to constitute social behaviour. The earwigs and certain beetles – dung and burying beetles, ambrosia and bark beetles, for example – have advanced quite a few steps further along the road to a social society. The earwig lays her eggs in a small hole which she digs in the ground, sits with them and broods them, turning them over and licking them from time to time. If she is removed, the eggs invariably fail to hatch: evidently her brooding is of vital importance. The young earwigs stay with their mother for only a few days after hatching: no true family life develops. In one species of burying beetle the female feeds her young on regurgitated food until the final moult. In some species of ambrosia beetles and bark beetles the females also feed their young. It is a prerequisite for the development of a society that the young shall possess a 'meaning' for the parents. There can be no doubt that their young have meaning for these earwigs and beetles. These earwig and beetle families are remarkable while they last. But they never last long enough to become societies: never beyond the one generation. These are what are known as 'sub-social' insects.

It is evident, therefore, that the elaborate social organisations of certain insects can be achieved only when the female lives long enough to have contact with several generations of her young. But, as Dr O. W. Richards points out in his *The Social Insects*, 'an even more important step is for more than one female to co-operate in looking after the young. This happens in none of the elementary social groups; it provides the best distinction between social and sub-social insects. A true social insect may be defined as one in which the female tends or helps to construct a brood-chamber for an egg (or larva) laid by another female.' This condition has been realised only in the order Hymenoptera and in the order Isoptera (the termites): in the most advanced of insect orders and in one of the most primitive. Since no termites occur in Britain we shall ignore them. But I should perhaps point out that they are unlike the Hymenoptera in that the males play as big a part in the daily life and organisation of the community as do the females. Insect-wise, this is a very primitive situation indeed. Only a very few insect species have developed fully social habits. And, outside the tropics, only the ants have achieved permanent social communities. In all the others the community is annual.

The bumblebees are social insects forming communities consisting of one female (the queen), a number of sterile workers – workers are undeveloped females, considerably smaller than full females but with larger brains – and, later in the year, a number of males and females. Only the fertilised females survive the winter, hibernating behind the loose bark of trees, in cavities at the roots of trees, or in mouseholes. With the coming of spring they wake up and set about the business of founding a family. Bumblebees differ in their habits. Some species are underground nesters, some (often called 'carder-bees') build their nests on the surface: but all begin the year in the same way, by searching for a suitable nesting-site and then building a nest. The carder-bees prefer thick grass or ivy (in my garden they have over the years shown a distinct preference for periwinkle): the underground nesters, being unable to bore their own holes, must find a ready-made hole and are particularly partial to the old nests of field-mice. (The necessity to find a ready-made hole makes it fairly easy to attract bumblebees. All you need to do is to bore a hole about a foot deep and with the shaft running at an angle of about 30 degrees, and to leave some nesting material nearby. Such holes are certain at least to be explored.) A suitable site having been found and the nest built – the nests are usually built of dead grasses and moss woven together (but all sorts of material may be used) and look not unlike the nest of a small bird – the queen builds a small spherical pot about the size of a hazel nut of wax and pollen.

In this she lays about a dozen eggs and then seals it up. She then builds a whole series of similar pots; some for the reception of more eggs, some for the storage of honey. The first bees make their appearance about three weeks after the

Female Sand Wasp dragging a paralysed sawfly larva to her nest

Female Leaf-cutter Bee on rose leaf

Female Mining Bee

Female Bumblebee (*Bombus agrorum*) on sallow catkins

laying of the first eggs. These first bees are all workers and they immediately begin to help their mother, who, up to now, has been entirely on her own doing the whole work of the nest. As more workers appear, so the queen is relieved of her tasks until finally she is able to devote the whole of her time to laying eggs. Towards the end of the season males and females are born and just before the end of the season (which is as early as the end of July with one or two species) these fly out and mate. The fertilised females then seek some secluded spot for hibernation. Their work completed, the males die and, as the evenings chill and draw in, so do the workers. The little community has reached its end.

The life of a community of social wasps is essentially similar. The young fertilised females hibernate in crevices in the trunks of trees and in nooks and crannies in walls. They also frequently find good resting places in houses and may often be found hanging by their jaws in the folds of curtains. The first action of the female upon waking from her winter's sleep, and this does not happen until fairly late in the spring, is to feed on the nectar of some plant. (Like the bees, all adult wasps are vegetarians, feeding upon nectar, fruit and fruit juices. Their larvae, however, unlike those of bees, are carnivorous and are fed upon insects captured and chewed up for them by the adults.) As soon as she has had a good meal the female sets about finding a suitable site for her nest. We have seven species of social wasps in Britain and they are far from particular about the choice of nesting site. While the common wasp generally prefers a hole in a bank or under the roots of a tree or shrub, she is not averse to making her nest in an attic. While the tree wasp generally lives up to her name, she may well take over a hole in the ground. If a hole is chosen, then the queen has to excavate a cavity at the end of it large enough to contain the first comb of her nest. This means carrying away the soil before building can commence: a long and laborious job. Indeed, until the first workers appear, which is about four weeks after the first eggs are laid, life for a queen wasp is terribly hard.

The nests of social wasps are wholly different from those of bumblebees and honeybees. The wasps have no wax. Their nests are built of paper, true wood-pulp paper. The queen scrapes shavings of wood from fences and similar places and makes them into paper with her own saliva. She starts her nest as a flat sheet of paper with a number of hexagonal cells hanging down from it. She lays one egg in each cell. When the larvae hatch out she has to forage for them, catching countless caterpillars and flies and other insects which she chews up and feeds to them. As her larvae grow she enlarges their cells, excavates more soil, starts the building of new cells, lays more eggs. Only when her first workers have emerged can she rest from foraging and building and devote herself to egg-laying. Now the workers take over the tasks of foraging and building and the nest begins to grow apace.

Stages of paper 'comb' are built below the first one, each hanging from the one above and so forming horizontal platforms along which the wasps can walk. Each layer or stage of comb, as it is formed, is surrounded by a strong paper envelope, an entrance hole being left at the bottom. This envelope is constantly being torn down, re-pulped, and used again as the nest increases in size. As it increases in size – and it will be the size of a football by the beginning of August – so the cavity in which it is hung must be enlarged to accommodate it. Every particle of soil dug out for this purpose is carried away and dropped a foot or more outside the hole. Only stones too heavy for a wasp to lift – and a wasp seems to be able to cope with anything up to twice its own weight – are left and they, of course, drop to the bottom of the cavity as it grows and so are nicely out of the way.

About the end of July or early in August the final stage of comb is built. This has larger cells, especially designed for the production of males and females. These take the normal four weeks to emerge from the nest. They mate and the females immediately seek secluded places for hibernation. The males die, but the community – it may number 2,000 or more by now – continues for a little while. But the year has turned. Insect food for the larvae becomes harder and harder to find and nectar is not so plentiful. The workers seem to realise that they can no longer function properly, that the community has no future. They destroy the cells with unhatched eggs: they cease to feed the larvae, leaving them to die. And soon they too, and the old queen, die of cold and starvation. All the skill in nest-building (and it is a superb skill, far exceeding that of any of the bees), all the months of ceaseless devoted labour, all ends with the production of a few fertilised females who will survive the winter to tread the same short road the following year.

As we have seen, insects can only become fully 'social' when the females live long enough to have contact with several generations of their young; only when the young help with the management of the nest and with the care and upbringing of succeeding generations. We have seen that the females of bumblebees and wasps are able to establish flourishing communities, but that these cannot survive the winter. They are summer communities only.

It is easy enough to understand why the wasps have not

been able to establish permanent communities. They are unable to make honey and their larvae have to be fed upon insects. With the approach of winter insects become less and less plentiful and wasps are unable to store food. The community is doomed to die of starvation. In the tropics, where at no season of the year is there any shortage of insects to provide food for the larvae or of nectar for the adults, there are permanent communities of wasps, which establish new communities by swarming; somewhat after the manner of honeybees.

It is less immediately obvious why the bumblebees have not been able to establish permanent communities. At first sight it would seem that all the factors which have played a part in the success of the honeybee are present also for the bumblebees. Both produce wax, both collect pollen, both make and store honey. Yet it is an unusually successful bumblebee community that numbers 400 individuals at the height of the season, whereas a perfectly normal community of honeybees will number some 50,000 individuals at the peak of its strength in early summer. Moreover, as we have seen, the bumblebee queen lives and heads her little community for but the one summer, whereas the honeybee queen lives and heads her community for two or three years (during which time she will lay as many as 600,000 eggs) and there are many authentic records of queens living longer than that. No doubt many factors help to account for this great difference in longevity and performance. Undoubtedly the most important is climate. Tropical bumblebees, like tropical wasps, do establish permanent communities; and new ones by swarming.

But if the British climate has prevented the establishment of permanent communities by the social wasps and the bumblebees, why, you may well ask, has it not also prevented their establishment by the honeybee? The answer is that the circumstances are altogether different.

The honeybee is often described as a domesticated insect. This is wholly inaccurate. The honeybee is a wild bee. Man has never succeeded in domesticating it in the way he has domesticated some other animals. All he has been able to do is to induce it to build its nest in some conveniently situated artificial site so that he may the more easily rob it of its honey. The honeybee is a native of India, Ceylon and Malaysia. In the wild it builds its nest in hollow trees and similar situations. But it has always shown itself eager to take advantage of any hollow site, from an earthenware pot to a wooden box, provided by man and man has been providing such artificial sites for centuries. The honeybee has also shown itself to be a most adaptable creature, readily taking to the most diverse sorts of flowers as sources of food. Man has, therefore, been able to disperse the honeybee all over the world. But the honeybee remains essentially a tropical insect and its permanent communities outside tropical regions survive only when they are housed in hives. Very occasionally a swarm escapes and establishes a nest in a hollow tree in this country. But I know of no authentic record of such a community surviving the British winter. The honeybee has, in fact, been no more successful in defeating the British climate than the bumblebee. Man has done that for it by providing virtually weather-proof hives.

The honeybee community is the highest form of social development attained by bees. There is the same division into castes – queens, workers, males – as in the bumblebees, but the differentiation between them is much more marked in the honeybee. The queen is easily recognisable by her long hind-body, which extends beyond her closed wings. The males (drones) are larger and stouter than the workers, with broader heads and much larger eyes. They are, moreover, stingless. The workers are the only members of the community – they form the vast majority of the community, of course – equipped to gather pollen, the only members of the community with honey-making and wax-secreting organs. They have, moreover, a much larger brain than the drones or the queen and they are armed with a formidable sting.

Queen bumblebee and queen wasp start their nests themselves, build the first cells, tend the first young. Not so the queen honeybee. She does not help in the life of the nest in any way at any time. She gathers no pollen, she builds no cells, she feeds no young. She is, in fact, no more than a highly-specialised egg-laying machine. She has no other function. For in the honeybee new nests are founded by division; by swarming. This happens, usually in May, when the nest (it is generally supposed) becomes overcrowded. No doubt the nest, the hive, is overcrowded at the time of swarming. But overcrowding surely cannot be the main reason for swarming. The decision to swarm must, in fact, be taken long before there is actual overcrowding, for it is the *old* queen who, with a few drones and a mass of workers (perhaps as many as half the total population), swarms from the nest, leaving a young virgin queen behind in her place: and it takes quite a long time to raise a young queen.

Queens are raised in special rounded cells which are much larger than the typical hexagonal comb-cells in which worker larvae are raised. Perhaps half-a-dozen of these special queen-cells will be built. The resident queen will be led round them by a band of workers and will lay an egg in each one. (These are, of course, fertilised eggs. There can be no doubt that the queen herself controls the sex of the eggs she lays by opening or keeping closed her sperm-containing sac. Presumably she knows from the size of the cell what is required of her.) The eggs take about three days to hatch and the larvae are then fed exclusively on a special substance, very rich in protein, which is known as 'royal jelly'. In about sixteen days the first of the new young queens will be ready to emerge from her cell. Shortly before she does so the old queen leaves with her swarm.

Young queens are apparently produced only when the old queen is getting past her work or when it is decided that a swarm shall be sent out. Obviously, at least in the latter case, there must be a good deal of advance planning. No insect has been more closely studied than the honeybee. There is a vast literature devoted to it and to bee-keeping. We know about its sight, about its powers of communicating information to other members of the community. It is evident that the honeybee community is thoroughly well organised, but we know absolutely nothing about its government. Always we are forced to use such impersonal, and unsatisfactory, phrases as 'it is decided'.

When the old queen leaves with her swarm of workers, each worker takes in its crop a ration of honey from which wax for the new nest will be made and with which life will be supported until new stores are gathered. Wherever the queen alights the swarm of workers will settle on top and all around here. It is at this point that the bee-keeper 'takes the swarm' and introduces it to a new hive, where the workers will at once proceed to build new cells. As soon as they are ready the queen will be led round them to begin her work of egg-laying. Egg-laying, by the way, is strictly regulated by the workers and can be increased or diminished according to the amount of food they allow her. The whole business is organised: but, again, by whom? The queen lays no eggs during the winter and during that season feeds herself from the storage cells.

The new virgin queen, on emerging from her cell, immediately makes her way to the other queen-cells and stings their occupants to death. This is her first duty. Occasionally, the workers will prevent her from murdering all the other queens, shielding one within her cell. This happens only when it has been decided (by whom?) that there shall be a second swarm later in the season. When this decision is made there can, of course, be no question of overcrowding, since half the population has just left with the old queen. If there is a second swarm the workers are accompanied by this second virgin queen, the first one, now fertilised, staying in the nest. Her second duty is to mate and she leaves the nest on her mating flight, pursued by a crowd of excited males. One mates with her in the air. She receives from him all his sperms (sufficient to last her lifetime) and in separating from him tears away part of his body, so that he dies. She then returns to the hive, to leave it only when she goes out with a swarm, to spend the rest of her life laying eggs.

When it is noticed that her egg-laying capability is failing, she is either killed by the workers or if there is a young queen available (and this is usually the case, for honeybee government is almost always far-seeing and prudent) no attempt is made to keep the two apart and the old is murdered by the young. With the approach of winter the food store has to be rationed, so the drones in the nest are maimed by the workers and then thrown out to die; and those outside are denied admittance and left to die of cold and starvation.

(This does not, of course, happen in honeybee communities in the tropics: a further testimony to the intelligence of the honeybee.) The life of the workers, the real rulers of the nest (or so it would appear) is not long. Those born in spring die in summer: those born in summer survive the winter (spending much of it asleep) and then live just long enough to rear the new spring brood.

CHAPTER SIX

THE TOP OF THE CLASS

Writing of the honey bee in his *An Insect Book for the Pocket*, Edmund Sandars has this to say:

> These bees and the ants, alone of our insects, have a permanent social community (which in the case of the bees we call 'the hive') with each individual doing its share of the duties involved in the community's continuance. Before giving the briefest review of the life of the hive, it is perhaps helpful to record the main differences between the communities of bees and ants. The bees' home in the wild state is built in a hollow tree, but under domestication in a hive designed by man to enable him to rob its contents without the need for killing all the bees. This he used to do throughout the ages until quite recently. The 'comb', with which the hive is furnished, is built of wax prepared in the bodies of the worker-bees, and has an architectural symmetry to which nothing in the ants' nest is comparable. Each young bee is reared from the egg to the adult stage in its own separate hexagonal tubular cell: the young ants are moved from communal ward to communal ward. Except for very brief periods, there is only one queen-bee in a hive at a time. The hive is stored with garnered foods: the ants' nest (of British ants at least) contains no stored food. Lastly, the bees fly out to gather food: the ants have to walk.

I think that anyone reading that, and having no knowledge of either insect, would be bound to come to the conclusion that in all important matters the honeybee has the edge on the ant. In fact, the ants are the most highly developed and the most successful of all the social insects. They are, in very truth, at the top of the class. But they are much more than that: they are also the dominant insects of the world.

Adult ants, as I have already pointed out, differ from all other insects in having a hump or tubercle (sometimes two) in the very slender waist. This gives the appearance of two (sometimes three) waists. You can never mistake an ant for anything else: nor, if you look closely, anything else for an ant.

As in the bees and the wasps there is a caste system: queens, males, and workers. The queens are generally considerably larger than the males; and the males usually, but not always, larger than the workers. It is impossible to be dogmatic about the relative size of males and workers because, in almost all species, there is an enormous variation in the size of the workers. The workers, which are always wingless, have much larger heads and much smaller eyes than the males and the queens. The workers often differ in form according to the work they have to do, which means that there may be more than three forms of a single species. The large head of the worker, by the way, contains a realtively large brain, greatly exceeding in size all the other nerve-ganglia throughout the body. The queen's eyes are smaller than those of the male. All ants have two stomachs, the first of which is the 'crop', the second the stomach proper. The crop may be considered the property of the community, the stomach that of the individual. Between them is a valve which, when opened, allows some of the food in the crop to pass to the stomach for the benefit of its owner. The contents of the crop, however, must be brought up for the benefit of the community, or members of it, upon demand. Some ants, queens and workers, possess stings connected to a poison gland. In other species, usually only in the workers, the sting is replaced by a mechanism which squirts the poison with quite remarkable accuracy over (for an ant) a considerable distance. Almost all British ants are 'squirters'.

So far I have tried to confine myself to British insects. Now, despite the fact that there are 41 different species of ants in Britain, I propose to give a brief account of ants as a whole; so intriguing and important are they. But before I do so, we should consider briefly the foundation and operation of a new nest, and for this purpose we will consider a very common black ant of gardens, *Lasius niger*. This is the ant which invades kitchens and larders and whose nuptial flight, usually in August, invariably gets a mention in the national Press. The flight always takes place on a hot, sultry day – the sort of day that makes one feel that there may be thunder in the offing – and usually in the late afternoon. The workers become very excited for an hour or two before the swarming occurs and take great care to prevent the winged individuals leaving the nest before the right moment. When the right moment does arrive – how is it recognised and by whom? – all the maiden queens in the nest and all the males are driven out by the workers. Since atmospheric conditions evidently control the moment of swarm, this happens to all the nests in the neighbourhood at the same time: so that in most years, somewhere, there are huge swarms of 'flying ants', large enough sometimes to disrupt traffic in cities. (H. St J. K. Donisthorpe records that on 8th August 1915 swarms of

Lasius niger – it was then known as *Acanthomyops niger* – were noted over the greater part of England from the Isle of Wight to Leicester and, though less plentifully, as far north as Newcastle-on-Tyne between 4.30 and 6 p.m.) Mating takes place in the air during flight and it is obvious that mating between individuals from different communities must often occur.

During mating the female takes all his sperms from the male (who dies, his sole purpose in life having been accomplished) and stores them in a special organ in her body, the *receptaculum seminis*, which is fitted with a valve which enables her to control the sex of the eggs she lays for the rest of her life. Mating completed the female returns to earth and immediately sets about getting rid of her wings. This she does by twisting them backwards and forwards, rubbing them on the ground or against a twig or a stone, and pulling at them with her jaws. They are not very firmly attached – they have, after all, to perform only one flight – and they come away pretty easily. Once wingless, she makes haste to get below ground. This means that she either returns to her old nest, enters another one, or sets about founding a new nest of her own. Some nuptial flights are very short. It is by no means impossible, and it may not be altogether unusual, for the female to be mated when no more than a foot or two above the ground and perhaps no more than a foot or two from her home nest. When this occurs she will return to earth just by her front door so to speak, and she will be welcomed back by the workers. If she should happen to return to earth in the immediate neighbourhood of another nest of her species, the workers of that community will probably go out and gather her in. But more often than not she sets about founding a new community of her own.

For this purpose she must find a ready-made cavity in the ground or dig one for herself. Into this she crawls and then closes it up behind her. She lies hidden in this underground chamber – it is known as the claustral cell – never venturing forth, until her young are born. She makes no cells, as do the bees and wasps: she just deposits her eggs around her. These eggs are minute: and the larvae which hatch from them are quite helpless and have to be fed by her with saliva from her mouth until they pupate. (The 'ants' eggs' which are sold by pet shops for feeding certain birds and fish are actually ant pupae.) All this time – and it may be a matter of months from the time she left the nest on her nuptial flight – she takes no food. She has to draw upon the reserves of fat in her own body, upon the re-absorption of her useless wing muscles, and probably upon a few of her own eggs, to keep herself alive and to provide food for her young. At the right moment she tears open the pupal cases to enable her brood of ants to emerge. From that moment her life is changed. She has had a very hard time raising her family. Now she is fed and cosseted by her young. Now she need do no more work. Now she becomes a specialised egg-laying machine. It has been estimated that for the rest of her life – and the normal lifespan of a queen ant must be about ten years, for many have been known to live fifteen years and more and the workers are known to live for up to seven years – she will lay eggs at the rate of one every ten minutes. If you care to do the sum, you will find that that amounts to quite a lot of eggs!

The first brood is, of course, composed entirely of workers: incomplete females. These first workers are always small because they have not had enough to eat and are always short-lived, soon working themselves to death, because, having been reared on a starvation diet, they have little stamina. They have but one duty: to bring food, more and more food, into the nest so that the queen may lay more eggs, so that the larvae may grow more quickly and pupate the sooner. These first workers rarely, if ever, do any building: they have neither the time nor the energy. But by the time they have worn themselves out, there will be other workers, larger and more energetic; and then others and others, each better fed and better cared for. Soon the nest will be in full working order.

As the numbers grow, which they do at a great rate, so the nest must be enlarged. And this means building. A few species build nests of paper, very like those of the wasps except that they do not have individual cells, and a few tropical species build their nests high in the tops of forest trees. But the vast majority of ants are essentially terrestrial, building their nests on the ground or underground. All the terrestrial ants, except the driver ants of Africa and the legionary ants of South America, neither of which have permanent nests, like a good firm roof over their heads. They have a particular fondness for flat stones – the paved terrace outside my house conceals a multitude of nests – but fallen logs, old tree stumps, anything that will make a roof is used. If a ready-made roof is not available, then a mound will often be raised above the nest. Ant hills are familiar sights throughout the countryside. Edmund Sandars pointed out that the architectural symmetry that lends such distinction to the nest of the honeybee is wholly absent from the nests of ants. It is true that ants' nests are in many respects simpler than those of bees and wasps: but this is an advantage. The bees and the wasps build their nests to an invariable fixed pattern: the ants have no rigid pattern, but build their nests to suit the physical conditions of the site. Ants' nests are simpler than those of bees and wasps in that they consist not of a great many individual cells but of a number of chambers, built at different levels and connected by passages. These chambers, which are of varying shapes and sizes, are used at different times of the day and at different times of the year according to the temperature and the humidity.

These chambers are used for different purposes. The harvesting ants – the ants to which Solomon directed the attention of the sluggard – use some as storage rooms. The honey-pot ants build some chambers with special vaulted roofs from which the replete ants, looking like swollen water-bottles, hang patiently waiting to be 'milked'. The leaf-cutting ants, the famous parasol ants of South America, the genuine fungus-growers, use some of their chambers in much the same way as we use mushroom beds. Other species, including one British, use some chambers as stables in which they keep root-feeding aphids. But the chief use of these underground chambers is as nurseries.

Workers carry the ceaseless stream of eggs to small chambers specially built for them, where they will get the right degree of warmth and humidity. When the larvae hatch, they are moved from one chamber to another according to the temperature; whether the day be hot or cold, wet or dry, always they can be moved to a chamber where the temperature will be just right. The honeybee larva grows in its beautiful hexagonal cell, shut off from its kind: the ant larva is hauled from chamber to chamber, dumped on the floor, hauled back again. Personal contact is the very essence of the whole exercise. The larval ant has 'meaning' for its nursemaid: some benefit accrues to the

Worker Honeybee on daisy

Queen Wasp

Head of Queen Hornet . . . greatly enlarged

Garden Black Ants (*Lasius niger*) with cocoons

nursemaid from the association. The larva produces a sweetish salivary secretion which is eagerly sought by the workers. Much of the care lavished on their larvae by ants undoubtedly owes its origin to the desire on the part of the workers for the secretion which they receive in return from the larvae. Some workers become professional nursemaids and spend the whole of their lives caring for the larvae and being rewarded by them. But, in general, work in the community is shared out; workers changing their jobs fairly regularly. After a spell as builder or repairer, the worker may have a spell in the nurseries, then a spell as cleaner – ants are very particular about the cleanliness of their nests – then a spell as forager, and so on. Whatever the job, that job is full-time. The nursemaid cannot leave her charges and go out for a breath of air, the builder cannot leave the nest to get some food. The food is brought in by the foragers, their crops full. They go round the nest, very much in the manner of a tea-trolley in a factory. The food passes from mouth to mouth; the hungry feeding from the full.

And what do ants eat? Certainly some ants are vegetarian: certainly some ants are carnivorous. The vast majority are probably omnivorous. I think that it is wiser to think of ants in terms of their occupations rather than strictly in terms of their diet: to classify them as hunters or food-gatherers or agriculturists. The hunters catch and kill insects and other small animals. These are usually dragged in one piece to the nest and there are cut up and divided among the inhabitants. The food-gatherers collect seeds, the nectar from flowers, and liquid excreta from aphids and scale-insects. The true agriculturists are the leaf-cutting fungus growers. But few ants are specialists in one occupation only. There is normally a good deal of overlap between the categories. Though the food-gatherers and the agriculturists never seem to take to hunting – they are, of course, not properly equipped to do so – some of the food-gatherers, and even some of the hunters, can be classed as agriculturists. It depends, I suppose, upon how strictly you define the term 'agriculturist'.

Two hundred years ago Linnaeus described aphids as 'ant-cows'. A hundred years ago Sir John Lubbock (Lord Avebury) showed that our common yellow meadow ant collects the eggs of the bean aphid in the autumn, stores them underground through the winter, and in the spring puts them out again on the food plant; thereby ensuring a 'dairy' herd for the summer. This can only be described as careful and far-sighted husbandry. This ant, which is the ant responsible for the grass-covered mounds (often thought to be molehills) in uncultivated fields, also keeps aphids which feed on the roots of plants. Our very common black garden ant herds its aphids above ground and – alone, I believe, among British ants – sometimes builds 'barns' of soil particles to restrain them and to shelter them from the weather. This ant will collect aphids and place them on plants near its nest, so that it can milk them with the minimum of inconvenience. These two ants can surely be classed as agriculturists, since they herd and tend their aphids very much as do men their cattle.

The wood ant, which is the largest species in Britain, builds large mounds of pine needles, small sticks and so forth, which sometimes reach a height of three feet or more, over their nests. The wood ant is a ruthless hunter, which destroys a large proportion of all other insects in the neighbourhood of its nest. Oddly, for so successful a hunter, the wood ant has no sting. But it can squirt its poison with remarkable accuracy to a distance of twelve inches, and so is able to immobilise its prey before the unfortunate creature has realised its danger. The wood ant, however, is not solely a hunter. It also keeps a large number of aphids on trees around its nest and also some in chambers in the nest. These are milked, but they are also killed for their meat; which is not the practice of either the meadow ant or the garden ant. Perhaps the wood ant cannot qualify for the title of agriculturist: but it cannot, I think, be denied that of rancher.

Many species of ants milk aphids: not all herd them. The jet-black ant *Lasius fulginosus*, which is common enough in the neighbourhood in which I live, nests in hollow trees and tree stumps and deep down among tree roots. It is commonly regarded as a 'dairying ant'. Certainly it milks aphids. But my own observation has not led me to believe that it keeps aphids in its nests nor have I seen anything to suggest deliberate herding on trees and shrubs. I think that it makes use of any aphid it comes across, just as it will make use of any carrion or seed that it happens to come across, and I do not think that this qualifies it as an agriculturist. I am, however, not so sure that it may not qualify in another way. Its chambers and galleries in hollow trees and tree stumps are strengthened with carton made of chewed fragments of wood and bark, which dries hard and looks not unlike the paper manufactured by wasps. This carton material supports a fungus. I think that there can be very little doubt that this fungus is deliberately introduced by the ants. As yet, it is not known whether it plays any part in their diet. If it is shown to do so, then *Lasius fulginosus* must surely qualify as an agriculturist.

Many species collect seeds. Our common black garden ant does so and so does our red ant. The real specialists at seed-collecting are the harvesting ants, which are found in many of the drier parts of the world. These ants live mainly on seeds, especially the seeds of grasses, which are stored in special granary chambers deep underground. In most species the seeds are brought in enclosed in the chaff. The chaff is removed inside the nest and then brought out again and thrown away on to a rubbish heap which forms a rough circle round the entrance to the nest. Most species, before storing the seed, bite off that part called the radicle to prevent it from germinating. Some species do not eat the seed in its natural state, but use it to make what is known as 'ant bread'. This is done by a special caste of workers, known as soldiers, which are larger than the normal workers and have enormous heads. The soldiers with their powerful jaws act as seed-crushers. The crushed seeds are then chewed so that the starch becomes mixed with saliva and is converted into sugar. Ant bread can be stored and keeps well. But most species store the seed.

It has been well established – indeed, this was known to the ancient Greeks and Romans, and probably also to Solomon – that if the seeds get damp in their underground granaries, the ants will bring them up and spread them out in the sun to dry. When they have dried, they will carry them all in again. This postulates a more than ordinary degree of organisation within the nest. It has been suggested that some species of harvesting ants actually cultivate the grasses they most appreciate in the immediate vicinity of the nest. The scientists appear to be more than a little horrified by the implications of this suggestion. Some have been at pains to point out that the fact that these grasses often grow thickly around a nest is not due to forethought but rather to inefficiency on the part of the ants. It is said that quite a lot of seed is accidentally dropped by the ants as they hurry back to the nest and that it is this seed which later germinates. A fortunate accident, no doubt: but still an accident.

Others maintain that the stored seeds do sometimes germinate and that these seeds are then brought out and thrown away on the rubbish heap. There they grow and the following year the crop of seeds thus produced will be harvested. But, say the scientists, there is no reason to suppose that all this is anything more than an accident. The theory that seed is dropped accidentally by the ants as they hurry back to the nest simply will not hold water. The trackways of these ants to and from the nest are often as well marked as much used roads. If these ants are so careless, then one would expect seeds to be dropped all along these trackways, the routes to be defined by grasses. They are not. The grasses grow only in the immediate vicinity of the nest. It would seem, therefore, that part at least of the second explanation must be correct: that seeds germinating in store are brought out of the nest. That is a deliberate act. Whether all the rest is accidental is a matter of opinion.

Slave-keeping is chracteristic of some species of ants, including our own *Formica sanguinea*, the blood-red or robber ant. When a young queen of this species founds a new colony she invades a small nest of another species, usually *Formica fusca*, the negro ant, which is an active friendly creature. She kills the fusca queen and in due course the workers, but she tends the fusca brood the while she lays her own eggs. When the new young fusca workers emerge they look after the sanguinea brood. They have become slaves, serving an aristocracy. The population of slave workers is maintained by means of regular raids on fusca nests. I was once, in the New Forest, so fortunate as to witness the whole of one such raid. It took place at six o'clock in the morning and was carried out with the precision of a military operation under a great commander. The fusca nest was surrounded by the main body of sanguinea troops and then attacked simultaneously from front and rear. In a very short while sanguinea workers emerged carrying pupae – they are said to take larvae as well, but I saw very few larvae being moved – and forming an orderly procession back to their own nest.

Sanguinea workers are perfectly well able to do all the work of the nest and, indeed, do do so should there be insufficient slaves at a particular moment. But they prefer to leave the domestic chores to their fusca slaves and to concentrate on the more exciting tasks, such as foraging, herding and raiding. There is, of course, a risk attached to slave-keeping: the risk of leaving everything to the slaves. There can be little doubt that this is what happened in the case of the amazon ants. In these ants the jaws are now unsuited for anything except fighting. Amazon ants are incapable of building a nest, incapable of looking after their young, and have such short 'tongues' that they are almost incapable of feeding themselves. They are very much larger and infinitely more formidable than their fusca slaves, but they have to all intents and purposes been conquered by them. The fusca workers build the nest and keep it in repair, forage for food, tend the brood, and feed the amazons: a very different state of affairs from that existing in a sanguinea nest, which they are never permitted to leave except for the removal of rubbish. The amazons could not now exist without their slaves. Clearly, in human terms, they are degenerate. But when it comes to making a slave raid they become formidable fighting machines, active and well organised, and have absolutely no difficulty in securing all the fusca pupae they require. Amazon ants of various species are to be found from France, right across central Asia to Japan, and again in North America. They must be considered pretty successful degenerates.

Apart from the large assortment of 'dairy cows' of different sorts kept by many species, the nests of ants provide a home for a positively astonishing variety of insects (I believe that I am correct in saying that more than 5,000 different species have been recorded from ants' nests throughout the world) which are generally referred to as 'guests'. Some, of course, are unwanted guests. There are, for example, a number of extremely unpleasant beetles which invade ants' nests, and take up temporary residence, for the sole purpose of killing and eating the brood or even, in a few cases, the workers. And there are a number of parasites, such as mites, which must be uncomfortable bedfellows and may be dangerous. Many other guests are scavengers, which are tolerated because they are useful. Among these must be mentioned the woodlouse. There can be very few ants' nests without their quota of woodlice.

The most successful of these guests have developed special secretions which are extremely attractive to ants. The caterpillars of some of the blue butterflies have a gland which produces a sweet secretion which is irresistible and which ensures that they are carried into the nest, where they can be milked at leisure. Perhaps the best known of these is the caterpillar of the Large Blue Butterfly, which is taken into the nest of *Myrmica ruginodis*, our red ant. The caterpillar rewards its host by feeding off its larvae. Addiction to this secretion is as dangerous socially for the ant as is drug addiction in man.

A number of small beetles are found in ants' nests all over

the world. Most of them have tufts of yellow hair which, it is thought, produce an aromatic substance pleasing to the ant. Certainly the ants lick these hairs and they take as much care of the little beetles as they do of their own young, often carrying them about the nest with them. There can be no question that they are really welcome guests and, provided that they do not become too numerous, they do not seem to do any great harm. And, in their case, there does not seem to be any question of drug addiction. A French observer has said that the licking of the hairs reminds him of a lady stroking her lap dog. It may well be that ants also have pets.

CHAPTER SEVEN

THEIR WORLD AND OURS

It should not be necessary to point out that the world of the insect, though it is part of our world, is very different from our world. It should not be necessary to point out that to a creature as small as an insect, the world presents problems to be surmounted and opportunities to be grasped which are, which must be, almost totally different from the problems and the opportunities which confront us. It should not be necessary to point out these things because, when we come to give the matter a moment's thought, it is all so obvious. But it is necessary that they should be pointed out because few people, even though they know all this, really appreciate the different dimensions of the two worlds.

I have a garden path which, because I like to see clean gravel, I keep as free of weeds as I possibly can. I find it pleasant to walk down this path on a summer day. I never think of it as anything but a path. But on a summer day this path is a forbidding desert to many a small insect. Similarly, a puddle no more than an inch deep may be a dangerous lake to many an insect. Twenty-four hours of drying wind in summer can mean drought and death to many thousands of leaf-hoppers. We cannot really think in these terms: they are wholly unnatural to us. I walk across my lawn. How many insects of whom I am wholly unaware, which I never so much as see, being six feet above them, do I crush to death on that short walk? A meadow of tall grasses must, surely, be a deep and dangerous forest for many terrestrial insects. But what do they think of my mown lawn? What do they think of my mowing-machine? What do they think of my big boots?

Now, I am being anthropomorphic. I am treating of insects in human terms. And that is a very deadly sin in the eyes of most scientists. But let me quote Dr Eltringham, F.R.S., a very great entomologist:

> In approaching the study of insects we must endeavour to free ourselves from the purely anthropomorphic standpoint. Living creatures, no matter how simple or elaborate their organisation, exhibit reactions to their surroundings. Such reactions are essential to their existence, not only for the purpose of avoiding dangers, and conditions inimical to their welfare, but also as a provision for their nutrition and reproduction. The simplest animals will frequently respond positively or negatively to such influences as light, heat, chemical substances, or physical contact. Such simple responses are known as tropisms, and whilst a positive tropism denotes a tendency to move towards the source of the stimulus, a negative tropism does not mean an absence of response, but a tendency to move away from the exciting influence. Thus insects which are attracted or repelled by light are respectively described as positively or negatively phototropic. Responses to what we should call tastes or odours, that is, chemical influences acting either by contact or at a distance, are referred to as chemotropic.
>
> Our own senses are so inevitably complicated by factors of consciousness such as pleasure, disgust, or aesthetic appreciation of what appeals to us as beautiful,

and many other mental modifications, that we find a difficulty in discussing the senses possessed by insects, since the words we use to describe human sensations may tend to convey a false impression of the nature of the responses observed in animals of a totally different organisation.

In vertebrates other than ourselves, whose organisation is nevertheless very similar, the psychic response to sensory stimuli is already far removed from human conception. How much further must it be in creatures whose nervous equipment presents hardly anything in common with our own. Provided that we entirely realise that the insect's nervous system differs widely from ours both in structure and probably in functional response, it is not, however, necessary or even desirable to avoid all use of terms commonly applied to our own faculties, merely because such terms are not completely appropriate. Insects, in common with other animals, are sensitive to external influences, and if their organs differ from ours, at least the influences are the same. We are at liberty to assert that an insect can see, without committing ourselves to any psychic hypothesis as to what it can or does perceive. We may further conclude that an insect can taste or smell, to the extent that it responds to contact or distant chemical stimuli, without implying the belief or assertion that it has the mentally aesthetic appreciation of the gourmet. Indeed, in the writer's opinion the reaction from the anthropomorphic standpoint in recent years has been somewhat over-exploited. As Minnich has observed, the gradual increase in our knowledge of the chemical senses of insects, and our consequent ability to compare them with the vertebrates, has resulted in the discovery of many rather fundamental similarities. I am inclined to venture even further, and to suggest, that the insect, highly specialised organism as it is, may, and probably does, possess a kind of elementary consciousness. Far removed as this may be from anything approaching our complex mentality, it may yet contain some trace of psychic emotion. I find it hard to believe that the butterflies assembled on Buddleia flowers, ripe plums, or other attractive objects, eagerly extracting the juices, fanning their wings in the sun and waving their sensitive antennae in all directions, are not experiencing something rather more than the mere unconscious adjustment of material appetite.

I, too, find it hard, impossibly hard, to believe. I think that Eltringham is absolutely right. He wrote that passage in 1932, when the anti-anthropomorphic movement was approaching its peak, at a time when we find another great entomologist writing this:

With the testimony of all these remarkable experiments before us, carried out by able and distinguished scientists, it would appear that there is no place for speculation on the intelligence of insects; that everything they do, no matter how intelligent the action may appear from our standpoint, must be the result of mechanical impulses; and that even if we observe certain actions which cannot be traced to any known tropism, it must be simply because that particular tropism which provokes that particular action has not yet been discovered.

I do not think that any scientist today would go quite as far as that. I would be very surprised to find that there is any scientist today who accepts whole-heartedly the law of mechanical impulses as the one law which governs all the surprising little ways, all the wonderful behaviour, of insects. Nevertheless, if one does not wish to be treated with disdain by the pedant, one must still be very careful about anthropomorphism; very careful not to interpret insect behaviour in human terms. The trouble about this is that I do not know any other terms by which to interpret it. I am well aware that when I use words like seeing, smelling and tasting, that when I speak of ideas such as liking and disliking, I am being anthropomorphic. But I know no other words to use. I am a human and I can think only in human terms. The fact that I use them does not necessarily mean that I automatically assume that insects' senses are affected precisely like our own.

Having made that clear, what do we know of the world these insects live in? What do they see? What sort of sense of smell have they? Is their hearing good or bad?

As I have already pointed out, the eyes of insects are of two quite different kinds; the compound or faceted eyes and the ocelli or simple eyes. It may be said at once that both are less efficient than the eye of the vertebrates. Neither kind has any means of focusing, which means that objects can only be seen distinctly or reasonably distinctly when they are close. The picture must blur, and rapidly, as the distance increases. Insects are short-sighted. But, owing to its peculiar structure, the compound eye is exceptionally well adapted for catching quick movements, as each facet in succession will be affected by a moving object. To the insect, perception of movement is much more important than that of form.

Can insects see colour? Yes. But their range of colour perception does not extend over the whole range of the spectrum visible to man: and it differs, of course, in different kinds of insects. As one would expect, most of the work on colour vision has been done with honeybees. The visual powers of the honeybee can be tested by a process of training. Small glass dishes, one containing sugar and water and the others water only, are placed on squares of coloured paper. The bees can be trained to associate the dish containing sugar with a particular colour. In this way it has been found that honeybees (and there is no reason to suppose that this does not apply to all bees) are colour-blind to red. Von Frisch found that his bees were unable to distinguish red cards from dark grey or even black cards. But he found that they could distinguish with a fair degree of accuracy between orange, yellow, green, violet and purple, and that they could distinguish any of these colours from grey of any shade. (A colour-blind animal, such as a dog, cannot distinguish a grey card from another colour of the same brightness.) Later experiments by Kuhn, while confirming all the results achieved by von Frisch, proved conclusively that bees see ultra-violet as a true colour and can be trained to it. If we think about this for a moment, we can see that the honeybee does not live in our sort of colour world at all. We can see red. The red flowers of the poppy are red to us. The bee cannot see red: yet the bee goes to the poppy to collect pollen. It has been shown that the poppy reflects a lot of ultra-violet light, which means that it appears blue or bluish to the bee. Again, some white objects reflect ultra-violet light: others absorb it. We see both as white. They will appear as two distinct colours to the bee.

And what sort of a sense of smell? That there is a sense of smell is beyond question, as has been proved by experiments with honeybees. In one such experiment a number of dishes are placed in boxes, which can be entered through a small hole. One dish contains sugar and water, the others water only. The one dish has added to it a trace of scent. The bees soon learn to associate sugar with the scent and will not then bother to enter unscented boxes. As one would expect, they are able to distinguish a great variety of scents and can detect scents in many things which appear to be odourless to us. The organs of smell are located on the antennae, which are extremely mobile and can be brought into contact with an object if necessary. Thus the faintest trace of scent will be discernible. It is by means of 'touch smell' that a bee is able to distinguish a member of her own nest from a bee from another nest.

So many kinds of insects produce sounds that it is logical to presume that they can hear. Some insects, such as crickets and grasshoppers, which produce a lot of sound themselves, have quite elaborate organs of hearing located in the abdomen or in the front legs. No such organs have as yet been discovered in the bee. This is really very odd because the first virgin queen to emerge from her cell will usually make a curious shrill sound, which is known as piping. Any other adult virgin queens in the hive, even though still imprisoned in their cells, promptly pipe in reply. There can be no doubt whatever that one queen hears the piping of another; though what she hears it with is not known. Worker bees seem to pay no attention to the piping of the queens and it has been suggested that they do not hear it. In fact, it is commonly believed that bees, despite the loud hum that they make, are deaf. That would not, of course, necessarily mean that they are insensitive to vibrations. The change from sound to touch is a gradual one. Unlike the bees, ants do have organs (chordotonal organs) suggestive of hearing in their legs. These organs have no tympana and a long series of careful experiments failed to produce any evidence of auditory powers. However, the ants did react to vibrations that came to them, not through the air, but through the earth of their nests and the wood and glass of the containers in which they were placed. This is very much what one would expect. Probably all subterranean animals are very sensitive to such vibrations.

The most remarkable advance in knowledge of honeybees was von Frisch's discovery of what he termed the 'language' of bees. I have not the space to deal with this fully here – details of it can be found in almost every modern book on insects and von Frisch's own book (which makes fascinating reading) is listed in the bibliography at the end of this book – but, briefly, what happens is that when a worker, having found an abundant supply of food, returns to the hive she communicates its whereabouts to other workers waiting in the hive. She does this by means of a dance, which conveys the information in a quite remarkably precise manner. If the food is no more than about 100 yards away, the bee performs a round dance; running round a circular track, first one way and then the other. This dance does not really tell the bees in the hive any more than that the food is not more than 100 yards away. That is a short distance. The bees in the hive will soon find the food, simply by quartering the ground.

If the food found by the worker is more than 100 yards away, a different dance is performed upon return to the hive. Now the bee runs along a track shaped like a figure of eight, but with the part where the two loops join straight. As the bee runs along the line between the two loops she waggles her abdomen. (The dance is commonly known as the 'tail-wagging' dance.) This dance gives a great deal of information, showing the direction as well as the distance of the food. The distance to the food is indicated by the speed of the dance: the further away the source of the food, the slower the dance, but the more wags of the tail on the straight part of the run.

The information given has been worked out very accurately: eleven figures of eight in 15 seconds for a distance of 130 yards, four figures of eight in 15 seconds for 2,000 yards. That is remarkable enough. Information about direction is even more remarkable. The dance is usually performed on the vertical face of the comb in the complete darkness of the hive. (It has, of course, been watched through the glass wall of an observation hive in a red light to which bees are not sensitive.) It is the straight run of the bee between the two loops of the figure of eight that indicates the direction of the food. If, on the straight run, the bee runs up the comb, then the source is towards the sun: if the straight run is downwards, then the source is away from the sun. If the straight run is at an angle to the right or left of the vertical, then the source of the food is at that same angle to the direction of the sun. But what happens if the sun is hidden by cloud? When the sky is completely overcast, no dance is performed. (Bees do not usually do much work when the sky is completely overcast anyway.) But if there is a patch of blue sky visible anywhere, then the tail-wagging dance will be performed as usual, the direction of the food being indicated in relation to the true position of the sun. Thus the bee must know the position of the sun, even though the sky is almost covered by cloud. And this means that bees are sensitive to the plane of polarisation of light and make use of the pattern of polarised light.

The development of communication in the honeybee – there seems to be nothing of the sort in the social wasps – is obviously an enormous advance in social organisation. Then what of the ants, which are socially, and in every other way, so much more advanced than the honeybee?

The trouble about ants is that they are not nearly so easy to study as honeybees. It is true that ants are much the easiest of the social insects to keep under observation in captivity. Many species can be kept successfully for years in small containers of plaster of Paris with a glass lid. All you have to do is to keep them damp and feed them on sugar and an occasional dead insect. Much of our knowledge of ants – of the early stages of a colony at any rate – is based on observation made on captive colonies. And that is just the trouble. They are captive colonies. They cannot increase. Their members cannot go out to work as does the honeybee. They are leading unnatural lives. We cannot, therefore, be certain that what we observe is natural behaviour.

This is not to say that we do not know a great deal about ants. We do. There is an immense literature on their natural history. And we know a great deal about their senses. We know that they have an uncanny sense of smell: more acute and more selective than that of the honeybee. We know that they have an uncanny sense of touch. We know that they follow scent trails and we know that, like the honeybee, they make use of the pattern of polarised light. And we know that they communicate with each other: that there is an ant 'language'. Indeed, there is a good deal of evidence to suggest that one species may have some knowledge of the language of another. It would seem evident that the slave-keeping ants must have some knowledge of the language of their slaves and vice versa. But it seems to go a bit beyond that. The common yellow meadow ant, for example, appears to have some understanding of the language of the common black garden ant and vice versa. Though (who knows?) it may be on no higher level than the understanding of French possessed by the average Public School boy, it does enable the two species to pass the time of day.

In a general way, I suppose, we know almost as much about the habits of ants as we do about the habits of honeybees. The difference is that we know not only what the honeybee does, but how it does it. We know that honeybees communicate with each other, we know what information is passed on and how accurately, and we know how this is done: by dancing, by tail-wagging. We know that ants also communicate with each other. We know that this is done – or, at any rate, that it is partly done – through the antennae. But that is all. We can (and do!) guess at much more. But we do not know. And this is because most ant species lead a subterranean, or largely subterranean, life.

And this, of course, is the reason for their great success. The ants have become dominant insects because they are wholly terrestrial in habit. William Morton Wheeler, the greatest of all authorities on ants, maintained that Espinas was the first (in 1877) to point this out and paid him tribute by quoting him at length.

Ants owe their superiority to their terrestrial life. This assertion may seem paradoxical, but consider the exceptional advantages afforded by a terrestrial medium to the development of their intellectual faculties, compared with an aerial medium! In the air there are long flights without obstacles, the vertiginous journeys far from real bodies, the instability, the wandering about, the endless forgetfulness of things and oneself. On the earth, on the contrary, there is not a movement that is not a contact and does not yield precise information, not a journey that fails to have some reminiscence; and as these journeys are determinate, it is inevitable that a portion of the ground incessantly traversed should be registered, together with its resources and its dangers, in the animal's imagination. Thus there results a closer and much more direct communication with the external world. To employ matter, moreover, is easier for a terrestrial than an aerial animal. When it is necessary to build, the latter must, like the bee, either secrete the substance of its nest or seek it at a distance, as does the bee when she collects propolis, or the wasp when she gathers material for her paper. The terrestrial animal has its building materials close at hand, and its architecture may be as varied as these materials. Ants, therefore, probably owe their social and industrial superiority to their habitat.

I have given that quotation in full because I consider it to be a magnificent piece of prose. Anthropomorphic? Of course it is: exquisitely anthropomorphic. And all the better for it. Moreover it happens to be true.

The honeybees are social insects which have been studied by man from the earliest times. They live in obviously well-ordered communities. But no human has ever drawn attention to, has ever seen, any resemblance between the honeybee society and the human society. We have the greatest admiration for the selfless industry of the honeybee. But that is as far as it goes.

Not so with the ants. It would be impossible to find two creatures less alike than ant and a man. Yet the resemblances between their societies are so conspicuous that they have been noted by man from the earliest times. I quote from Wheeler's classic work *Ants:*

> Folk-lore and primitive poetry and philosophy show the ants as an a biding source of similes expressing these fervid activity and co-operation of men. Although these similes have become trite from repetition, the scientific student can hardly free himself from the many anthropomorphisms which they suggest. He is forced to admit that the social and psychical ascendancy of the ants among invertebrates and of the mammals among vertebrates constitutes a very striking example of convergent development.

Sir John Lubbock, in his *Ants, Bees and Wasps* (1894), puts it like this:

> Whether there are differences in advancement within the limits of the same species or not, there are certainly considerable differences between the different species, and one may almost fancy that we can trace stages corresponding to the principal steps in the history of human development. I do not now refer to slave-making ants, which represent an abnormal, or perhaps only a temporary state of things, for slavery seems to tend in ants as in men to the degradation of those by whom it is adopted, and it is not impossible that the slave-making species will eventually find themselves unable to compete with those which are more self-dependent, and have reached a higher plane of civilization. But putting these slave-making ants on one side, we find in the different species

of ants different conditions of life, curiously answering to the earlier stages of human progress. For instance, some species, such as *Formica fusca*, live principally on the produce of the chase; for though they feed partially on the honey-dew of aphids, they have not domesticated these insects. These ants probably retain the habits once common to all ants. They resemble the lower races of men, who subsist mainly by hunting. Like them they frequent woods and wilds, live in comparatively small communities, as the instincts of collective action are but little developed among them. They hunt singly, and their battles are single combats, like those of the Homeric heroes. Such species as *Lasius flavus* represent a distinctly higher type of social life; they show more skill in architecture, may literally be said to have domesticated certain species of aphids, and may be compared to the pastoral stage of human progress – to the races which live on the products of their flocks and herds. Their communities are more numerous; they act much more in concert; their battles are not mere single combats, but they know how to act in combination. I am disposed to hazard the conjecture that they will gradually exterminate the mere hunting species, just as savages disappear before more advanced races. Lastly, the agricultural nations may be compared with the harvesting ants.

The analogy to the three great phases in the history of human development – the hunting, the pastoral, and the agricultural – is so striking that it cannot be ignored. Wheeler accepted it without question. But he went on to point out that there is no resemblance between the ant society and the human – that the ant society is composed of sterile females, each of which works for the good of the state alone: that there is no private enterprise – and that the ants have never taken a step towards the final three phases in the history of human development: the commercial, the industrial and the intellectual. And, of course, all that is perfectly true. Nevertheless, the resemblances are so strong – only the ants and men wage organised war: only the ants and men have succeeded in domesticating other animals – that for most humans they override everything else. You cannot think about ants without being anthropomorphic.

The battle to put the ants in their place, and keep them there, led some scientists, as we have seen, to treat them (and the bees) as mere reflex machines, operating at the dictate of tropisms. Von Frisch's discovery of the bees' dance has put paid to that and also to the miraculous powers of instinct in which Fabre was so firm a believer. It was reaction to the 'reflex machine' school that prompted Maeterlinck to write his books, ascribing the achievements of insects to intelligence alone and drawing some pretty invidious comparisons between their minds and that of man. The pendulum has been swinging back and forth ever since: which, if nothing else, is a striking tribute to the fascination of the social insects.

The resemblances between the two societies being so striking, men will inevitably continue to ignore the equally striking differences (though I have to admit that these differences have grown just a little less striking in the last fifty years; philosophically, at least), and will continue to compare them and in anthropomorphic terms. I think this is good. Anything that makes you think, no matter how violently you may disagree, is good. I would, therefore, like to leave you with a thought.

A recent attack on the anthropomorphic attitude to ants has pointed out that ants trapped in the Baltic amber thirty million years ago resemble in every detail, even to the mites on their legs, ants running about English gardens today. In other words ants ceased to evolve thirty million years ago. They are in a cul-de-sac, at a dead end. Man, it is pointed out, is still evolving, is marching down the broad highway of progress. But is evolution necessarily a continuing process for everything? Can it be that the ants attained perfection for their way of life thirty million years ago and – oh, how anthropomorphic can you get? – knew when to stop. Is it possible that still-evolving man is on the brink of evolving himself to extinction?

SUGGESTED FURTHER READING

BUTLER, C. G. (1954). *The World of the Honeybee*. Collins, London.

CARTHY, J. D. (1956). *Animal Navigation*. Allen & Unwin, London.

CROMPTON, J. (1948). *The Hunting Wasp*. Collins, London.

—— (1954). *Ways of the Ant*. Collins, London.

DONISTHORPE, H. St J. K. (1927) *British Ants*. Routledge, London.

—— (1927) *Guests of British Ants*. Routledge, London.

EWERS, H. H. (1927). *The Ant People*. Dodd, Mead & Co., New York.

FABRE, J. H. (1912). *Social Life in the Insect World*. Fisher Unwin, London.

FOREL, A. (1928). *The Social World of the Ants*. Putnam, London.

FRISCH, K. VON (1954). *The Dancing Bees*. Methuen, London.

GOETCH, W. (1957). *Ants*. University of Michigan Press, Ann Arbor.

HASKINS, C. P. (1945). *Of Ants and Men*. Allen & Unwin, London.

HUXLEY, J. (1930). *Ants*. Chatto & Windus, London.

IMMS, A. D. (1947). *Insect Natural History*. Collins, London.

WHEELER, W. M. (1910) *Ants. Columbia University Press*, New York.

—— (1928). *The Social Insects*. Kegan Paul, London.

INDEX

Abdomen, 14, 19, 21, 27, 30
Alimentary Canal, 20
Ant: agriculturists, 90, 91
  Amazon, 94
  black, 84, 85
  dairying, 91
  driver, 87
  flying, 84, 85
  harvesting, 88, 91
  honeypot, 88
  leaf-cutting, 88
  legionary, 87
  Nurseries, 88
  parasol, 88
  slavekeeping, 93, 109
  soldiers, 92
  'squirters', 84
  velvet, 31, 47
  wood, 90, 91
'Ant bread', 92
Ant language, 106, 107
Ants, 109
Ants, Bees and Wasps, 109
Antennae, 12, 14, 15, 51; ants, 107
Aphids, 58, 88, 90, 110

Balfour-Browne, Professor, 62

Bees: bumble, 64, 72, 74–7
  cuckoo, 31, 48, 49, 63
  honey, 74, 76, 77, 83, 87, 102, 104, 106–108
  leaf-cutter, 50, 56–8
  Melecta, 50
  mining, 50, 56, 58
  potter, 50, 56, 57
Blood-suckers, 15
Brain, 16, 22; ants, 83
Brood-chamber, 71
Bumblebees, 48
Butler, Colin G., 48

Carnivores, 56, 58, 73, 89
Cells, 58, 74, 75, 78; ants, 88; claustral, 86; Embryonic, 45
Central nervous system, 20–2
Cerci, 19
Chitin, 13, 14
Chordontal organs, 103
Circulatory system, 20
Cocoon, 28, 30, 66
Coleoptera, 25, 55
Comb, 74, 75, 82
Conidia, 29, 30

Diptera, 27, 55

INDEX

Donisthorpe, H. St J. K., 84
Drone, 78, 80, 83

'Ears', 11, 15
Egg-laying, 19, 34, 35, 36, 54, 73, 78–80; ants, 86
Eltringham, Dr, 98, 100
Eyes, 15, 78, 83; compound, 15–17, 101; facets, 15, 16, 101; simple, (ocelli), 16, 17, 101

Free, John B., 48

Galls, 27, 31ff., 40, 41, 56
   Aleppo, 39
   bean-gall, 34, 35
   formation, 37
   gall-flies, 34, 36
   gall-midges, 31
   marble-gall, 39, 43
   pseudocone-gall, 31
   purse-gall, 31
Genitalia, 19
Gerard, John, 33
Globular gaster, 48

Haldane, 23
Hearing organs, 15
Hibernation, 49, 72, 73, 75

Imms, A. D., 42
Inquilines, 40, 41, 43
Insect Book for the Pocket, An, 82
Isoptera, 54, 71

Keratin, 13

Labium *See* Mouth-parts
Labrum *See* Mouth-parts
Larvae, 26–30, 33–5, 37, 42, 45, 50–2, 55, 56, 58, 59, 62, 66, 71, 73–6, 79, 95; ants, 86–9
Larval secretion, 35
Legs: coxa, 18
   false, 28
   tarsus, 18
   tibia, 18
   trochanter, 18
   walking, 12
Lepidoptera, 25
Lubbock, Sir John, 90, 109

Malpighi, Marcello, 33
Mandibles *See* Mouth-parts
Mating, 19, 68, 75; ants, 85; flight, 80
Maxillae *See* Mouth-parts
McCallan, E., 42, 43
Mouth-parts, 14, 52
   Labium, 14, 26
   Labrum, 14
   Mandibles, 14, 26
   Maxillae, 14, 15
Muscles, 13, 19, 20, 22

Nests, 56ff., 74, 75, 77, 87, 88; ants, 87, 88, 92, 95
New Naturalist Library, The, 48

Omnivores, 89
Organs, location of, 11ff.
Ovipositer, 19, 27–30, 37, 39, 41, 44, 45, 51

INDEX

Parasites, 31, 34, 37, 41–53, 54, 55, 95; endo-parasites, 46, 47; exto-parasites, 46; ichneumon-flies, 31, 42, 50
Parthenogenesis, 37, 38, 45
Petiole, 48
Phylum Arthropoda, 12
Picard, F., 44
Pliny the Elder, 32
Pollen, 76, 78, 102
Pupae, 59, 62; ants, 86, 87, 94, 95

Queen, 72, 73–80, 82, 83, 103; ants, 84, 86

Receptaculum seminis, 85
Respiratory system, 20
Richards, Dr O. W., 71
'Royal jelly', 79

Sandars, Edmund, 82, 87
Sap-suckers, 15
Saw-flies, 26–8, 31, 34, 35, 38, 55
Sclerotin, 13, 14
Skeleton, 20; cuticular. 13; exoskeleton, 13, 14, 19
Social Insects, The, 71
Social insects, 68–81, 108
Solitary insects, 50, 56–7, 68, 70
Spiracles, 18
Stings, 54, 78; ants, 90
Stomach, 83; crop, 83, 84
Swarming, 77, 78, 80

Tail-wagging dance, 104, 105, 107, 111
Theophrastus, 32
Thorax, 14, 17, 19, 21, 27, 28, 30, 48; mesothorax, 17, 18; metathorax, 17, 18; prothorax, 17, 18
Tracheae, 12, 13, 18
Tropisms, 98
Tubercle, 83

Vegetarians, 56, 73, 89
Venom, 59
Vertebrates, 11, 19, 20

Wasp: biorrhiza, 36, 37, 40
chalcid, 31, 41, 44–7
cuckoo, 31, 50
cynipid, 31, 35, 40, 41, 43, 55, 56
digger, 48, 56
gall-, 31, 34–7
mason, 56
potter, 56, 58
rhodites, 38, 40–2
ruby-tailed, 50
sand, 56, 60, 61
wood-, 26–30, 51, 55
Wax, 76, 82
Wheeler, William Morton, 107, 109
Wings, 12, 17, 18, 24–6, 37; ant, 85; fore-, 25; hind-, 25–6
Workers, 72, 74, 77–9, 81–3, 103, 104; ant, 84–6, 89

36584

Vesey-Fitzgerald

The worlds of ants, bees and wasps

DATE DUE

MAY 2 1 1973

APR 16 '91

DISCARDED